CYBERBABY
('*KILL IT IN THE WOMB*')

A High-Thoughts Production
By Mark Johnson

eljohnsonwaynee@yahoo.com

For those with eyes to see…

1

December 16, 1996.

She stood between two dangers. One was real and the other was not but both were equally potent. There was a one hundred-foot drop in front of her! That was real, and Gyasi begged her not to go any further. The devil was coming to tear her apart! That was not real, but, as she perceived it as such, it counted as danger to her anyway.

She turned her back to the cliff, not wanting to jump. Her arms outstretched in confidence that Gyasi would rescue her. He coaxed, begging her to wait for him but it seemed like her invisible attackers were closing in. Lisa heard his voice, yet she stepped back unwittingly. Gyasi's heartbeat intensified, pounding mercilessly against his ribcage until it felt bruised. He made to call at her again but no words came, for his mouth was too dry.

She hesitated, which he took for a sign that he had gotten through to her. Then unexpectedly, she turned and jumped with his name on the tip of her tongue! He saw hell unfold but could not believe! He pleaded again belatedly, "Lisa! *Noooo!*"

She landed on her back, in a floral maternity dress that should have kept below the knees. Gyasi got to the steep edge just in time to see her take her last breath. The lights went from her hazel eyes. Pale ones stared up into his face, looking beyond him to an extra-dimensional plane. Proud Lisa lay sprawled upon solid rock, with no more care for decency. For the first time since the insert, her stomach seemed inflated, but it did not even matter anymore.

In the midst of the tragedy, Gyasi could not help wondering if she blamed him. He would be eternally debited!

"Dianne!" the horrified little man shrieked. The mercuric oriental peeped through the trees and saw from a distance. Instead of coming to him, she turned and headed back to the house. What happened was unchangeable! Dianne loved to think before she talked.

It was near sunset now. Golden beads of light beamed through the trees and sparkled against the sweat of his skin. Aduna still expected to awake from a horrible dream. He did not know how to turn and walk away! Ten minutes later, sunset came and turned to dusk. Still, he waited. Time passed, but she was dead anyway.

Then nature changed in an instant, to complement that eerie depressing mood. Gusty winds whirled forth, rocking the naked pine trees around his secluded property. They threatened to uproot and tear the leafy willows apart. Rain started to drizzle and snow came with it. He leaned his head to the left and poked his right eye at the sky, as if to accuse the Celestials that bedeviled him. The tolerable temperature transformed to freezing cold. What never happened in years happened in seconds. Things changed from good to worst!

He stood above the one hundred-foot drop, looking down on her dead body and bracing against the strong breeze. From the house to that ravine was over a hundred and fifty meters. The trees grew so thick that he could hardly make out the building. He saw Dianne's shadow against the dim light of his back porch, pushing forward through the undergrowth.

Aduna chose this place because it was private. It was twelve acres of real estate and he built the house exactly in the center. It felt safe out there, until that December evening.

The fault was with a program in the embryo. He should have been wise about it. Now he would learn his prudence for the price of innocence! A life was lost on his account! Aduna Gyasi felt the pain but was powerless to make amends. He gained her trust by assuring her that it was safe. That day he watched her die, arms outstretched in confidence, waiting for him to rescue her. He failed the mission, he failed the world and he failed Lisa. She died believing in him!

Gyasi looked down again. A boisterous wind blew his roomy white robe around his diminutive frame. The young scientist seemed peculiar to any ethnicity on earth. He was dark, with long white hair and beard, and fingers like eagle's talons. Unusual as he seemed, Aduna Gyasi knew how to hurt as much as any other soul on earth. It troubled him to think that he begged her to take the risk!

He was not a woman! He could not take on the burden of nurture and labor! It was his mission though. He should not have subjected her to chance. She was a good candidate. Healthy women who did not eat flesh were hard to find. It took months to convince her. What did he prove after all? If she refused, she would be alive! He stood on the edge reliving the tragedy repeatedly, like that was punishment enough!

Gyasi programmed the seed, from his own body, that impregnated her. He created a superior genome structure by natural and not artificial means. He designed biological programs that incited mutations in her womb. The god-man made her compatible for the peculiar hybrid life force. He could not have erred!

Somehow, a synthetic neural structure was catalyzed inside of Lisa. It was alien as far as he knew. This strange program leached into her system and took on its own volition. It became a virus, affecting her brain. The demonic parasite created illusions in her mind. She was no longer sure about reality. The one truth she trusted was that Aduna Gyasi had power to save her. Lisa believed that extraterrestrial legions came to destroy the project in her. She fled the invisible attacks from adverse thoughts, screaming for him.

The wizard took a glance at Dianne and turned back to face his woes. She was an oriental, always dressed to look the part. She stepped up beside him to peek. Although she resolved to strengthen Aduna, she covered her face and tensed unwittingly. Lisa rested at the center of the rock with blood streaming all around her. It appeared as if she was in the middle of a red spider's web. Dianne's compulsive reaction affected her partner.

"Oh my god!" the scientist lamented, "What did I do?"

The god-man clenched his fist, berating himself. Lisa Stare was a young and jovial person who had many years ahead of her. She came to him with awesome potential and love for science. He gave her death!

Dianne placed her arm across his shoulder affectionately, "What 'we' did, including her! If it were I down there, I wouldn't blame you! We shared the same dream. She didn't die in vain!"

Her show of nonchalance was mere pretense but her fortitude impressed him. The god creator leaned his head sideways to figure. "Why?" he agitated.

"Because the mission will move on and we'll be wiser."

"No way! I was wrong!"

"You're never wrong Aduna! Let the charge for her life be on me!"

"Nonsense!" he snapped.

"It will be, if the mission does not continue!"

"Why do you say that?"

"I'll tell you why! *Do to others as you'd have done to you*!"

"I could not do it to myself! I am male!"

"I'm woman and I'm yours! As you subjected Lisa, prove to the universe that it was without prejudice! Let me bear the cyber-babies and your seed grow in me – *your* wife! Not in another woman...a stranger to you!"

"Are you insane?"

"*No*! Are you a hypocrite? Truth has no regard for sentiments! You don't like to agree with it, but it remains unanswerable!"

"Look!" he fumed, "Lisa, your friend and mine is dead because of my insanity!"

"Your insanity will save Mankind! One life is precious to you and me! Since it is your project, must we prejudice the billions that you weigh your priorities against?"

She paused for him to take up her challenge. He remained silent. Dianne took a deep breath, assumed the role of authority and declared, "We'll mourn for today!

Tomorrow you'll prepare me to carry my burden that will save billions! My sacrifice will exonerate us in the name of the Law of Perpetual Motion!" She put her arms around him affectionately and added, "Come now! It is what it is! I came to take you home!"

By her determination and those words of conviction, Dianne Gyasi kept the dream for Mankind alive. The decision to continue with the mission meant that Lisa had to make another sacrifice, even in her death. They reported her missing. The project remained a secret.

He was exhausted that his eyes felt swollen. The little wizard tossed in bed, willing himself to sleep but the result was mere frustration!

His mind drifted back to December, when he did the insertion on Dianne. Gyasi tried to convince her not to travel until after she delivered, but he could not give her a logical reason. Dianne was not one to convince without logics. He bade her farewell and she went to Bombay with the children, leaving him to celebrate the New Year alone!

Gyasi smiled. He could not sleep and she was the reason! Dianne was good for his equilibrium. She was Libra by western calculation. Whoever said Capricorn was an unlikely match for her was wrong! It was the sign of his initiation to divinity. In this age and time, he came as stalwart of gods. The greatest man who ever was, or ever would be, had to survive Capricorn.

"Satan being!"

Gyasi's eyes popped open and he sat up in bed abruptly, his heart shocked into overdrive. It seemed that he had finally dozed off, but it did not last long. He heard someone! He turned to the bedside table for a remote to put the light on. Then he saw that he was elevated above his human form, which slept peacefully below him! Was he *dead*? "What the devil is…?"

"You are the devil Freeman!" the voice came back with malicious flavor. *"Welcome to 1997, the year of your possible demise!"*

"Who are you, and why would you destroy me?" the little scientist questioned, looking around anxiously to locate his company.

"I am here by your window!"

Gyasi turned to see the angelic being. It was a huge ivory colored humanoid entity and behind it, an oval-shaped craft hovered in the space of his window.

"Well?" the creature asked testily. *"Are you afraid?"*

"Why should I be?" Aduna questioned. "I choose how I want to feel. Now, I choose to feel curious."

The creature laughed and it sounded like a mare neighing. *"You are a free man! I will satisfy your curiosity! Come to the craft! Let's travel and we will talk!"*

"I am sleeping in my realm! If you want to talk, I will oblige you from here!"

The beast shrugged, *"I see you have great survival skills too!"* At that, it cackled before continuing. *"You are a newbie to the high thresholds, so let me tell you! I am a messenger from a realm of shepherds that watch over this mediocre domain! Call us 'the Watchers' if you please. My visit should count as an honor to you. If it turns out void of agreement, it shall prove to be your destruction! You cannot fight the children of the illuminated ones and win Aduna!"*

"We are from different realms. What is there to fight for? Are you a rogue of your kind? Would it not be wrongful interference on your part?"

"I know the divine laws more than you do ex-sheep! We understand our rights, and we work according to the treaties! Our influence is purely from above, by ethereal designs. There is no law against that! There is no law against the celestial sun passing over Saturn's realm, where Saturn's subjects abide in suffering. There is no law against the vitamin you receive from his passing, which the lord of this world did not give you! Neither will there be a law against those we influence by our legal spirits!"

"You have nice ways to propose unrighteousness! Chivalry is alive in you! It has been written in concrete from the beginning Watcher being. Satan will ever rule in Hades over the seeds of heaven for his appointed time. The great life-giver who resists Satan is Jesus the beneficent. He ever returns in spirit, to bear his seed to glory through the resurrection. Are you the Holy One of Heaven, or are you the imposter?"

"It seems you have a penchant for hurling insults!"

"Why would you consider my inquiry an insult?"

The herculean being ignored his question and said, *"Let me get to the point, you obnoxious spawn from mediocrity! We know about the Cyber-Baby Project. You want to seed the masses into higher evolution by crossbreeding them with superior genotype."*

"You know that?" Aduna Gyasi sounded surprised.

"We know all free man! We have watched you evolve! I am here to warn you of the detriments in your cyber-being idea. It is a threat to the Watchers' authority and an abomination to the gods. You must desist!"

"Why must it threaten the Watchers? Do you even speak on their behalf? It is an earthly affair. The gods speak for themselves! They speak through me every hour of the day. All my principles and designs come by their inspiration!"

"Fool! You do not understand! We will influence the evil humans to strike you where it hurts! As you love your family, the humans will destroy them! For everything you do, there will be a mountain to climb. Then there will be another one, greater than the one before! Feel very threatened until you relent!"

Aduna paused and then began unconvincingly, "You have no right to interfere! I…"

"You will interfere with us through your works! To that same degree, we shall interfere with you! We will not touch you with our hands or attack you by our persons, but persons shall handle you wickedly and persons shall attack you! No one and nowhere will be safe for you to trust or abide! The sons of man are our sheep! They will do our biding despite you imbecile!"

Suddenly, the room became cold, but he sweated! Fear took hold of Aduna! He heard his heart's rhythm, throbbing at the temple of his ears. The entity smiled knowingly, cackled and said, *"We will send you a warning! You will know by the sound of the echo!"* Then it disappeared.

He felt himself falling backward and panicked. Then he was in his body looking up at the ceiling, wondering if he could trust going back to sleep. He loved his family but the project was as life itself! Their images kept popping up in his mind to remind him. Was it karma? Would he betray them, as he betrayed Lisa?

....

January 3.

There was no parking in the pickup zone at Logan's Airport. They would have to stand outside until he drove around. The flight was due in at 1:45 PM. He got there at 2:15, hoping that he would not have to wait too long.

The minutes ticked by without a call from her! He shifted impatiently. "What time is it?" the wizard asked lazily. The sixth sense implant dialed up in his head. Dianne's voice informed him, *"The time is now 3:46 PM!"*

He loved that voice! When he made his personal implant, he decided to use it for his default. The sixth sense was handier than any laptop computer. Unfortunately, it would not hit the market for the next decade. Monocle and spectacle televisions, regular and contact lenses were old technologies in his catalogue. The masses did not know about them! He needed an investor to get his works out faster. It was not about the money. He made enough from what he managed to sell. Gyasi enjoyed seeing his handy creations benefit the people. He wanted them rolled out before they were outdated.

The sixth sense was the cutting edge of digital technology. He implanted it underneath the skin near his temple. He could make calls by simply touching his temple with his index finger and repeating the name or number. The device would go to directory, retrieve the digits and make the call. Incoming calls were voice activated, but the phone would suggest who was calling in advance. If he did not want to take the call, he would simply say 'no'. The sixth sense would tell the caller that he was unavailable.

"Dianne is on the line!" the sixth sense informed him.

"OK!" he responded eagerly, sounding up beat. The call came through.

"Hello? Aduna?"

"Yes dear! I am here!"

"We're waiting for you at arrivals now."

"Great! I am on my way! How was the trip?"

"Generally, very good! Thank you…although you didn't come! You disappointed my mother!"

"Do not worry dear. I know the way back to her heart!"

"I'm sure! She spoils you too much!" Dianne hung up.

Within a couple of minutes, the green Ford Explorer pulled up to the curve. Gyasi lowered his right window joked, "Look at you! All dressed like little Indians!" He hopped out of the car to help them load their stuff promptly but the siblings creamed him in a group hug. "My three little personalities!" the little man cooed affectionately. Dianne paused to watch. "They're happy to see you!" she said ironically. "Although you chose not to come, which is like dumping us!"

Gyasi fought his way out of their clutches and said, "You are not going to let me live this one down, are you?"

"Right now it seems unlikely," she quipped, "but I'll think about it for your benefit!"

He leaned his head sideways, watching her questioningly. Finally, she relented, "Alright! Just one little hug!"

As they embraced, he joked, "I smell curry in your hair!"

She put her mouth to his ear and whispered, "It was 'curried buck tongue'! You should smell the Bombay hair at the bottom!"

"You did not!"

"You did not come, so you do not know!"

Aduna eased her off him so they would clear the zone before the cops moved to hurry them along. "Get inside the car kids!" Dianne instructed.

Two security officers trudged toward them from forty feet away. One was a rugged big boned ginger-haired youth in his early twenties. He had a huge male German shepherd dog. The other was a Hispanic woman, walking twelve feet wide of her colleague.

Ray and Charm got inside the vehicle. Dawn held her place. She fixed her curious eyes on the animal. The three-year old had a white teddy bear clutched to her bosom. Then she remembered! She turned to Aduna who was still loading up. "I brought this for you Daddy!"

"You did!" the wizard replied excitedly, "Keep it for me until we get home!"

"Dawn!" Dianne butted in sternly, "Get inside the car!"

The child seemed oblivious to her mother's instruction. She watched the animal with intense curiosity. An albino looking officer walked up to the dog handler and whispered something to him. The youth nodded agreeably. Then the stranger put out his hand as if to touch the German shepherd. The dog wagged its tail in a friendly way.

Dianne's eyes tracked to Dawn's focal point. She smiled and informed her husband, "I think we'll have to get this kid a pup very soon!"

Aduna did not answer her. She watched her daughter watch the dog. Then the albino fellow turned and pointed directly at the child. The dog handler nodded. Dianne frowned, but it turned into an incredulous smile. Did they think she had drugs in her teddy bear?

The albino walked off and disappeared. Then the dog handler pointed to Dawn and shouted, "Watch that kid!"

Dianne frowned. Aduna pulled his head from inside the trunk confused, "Huh?"

"He's not talking to us!" Dianne surmised. "He couldn't be!"

Suddenly, the handler let the dog go and it took off for Dawn in a seething rage! His colleague stiffened in panic. "Watch that kid!" she called out hysterically, charging across to help Dawn. "Down C7!"

She was breathing heavily with fear, knowing it was futile. The canine would not take orders from her! Her ginger haired counterpart hissed in annoyance, "I told you to watch that kid! Bad kid! Bad kid!"

"Shut the fuck up Drake Shepherd!" the Hispanic woman screeched. "Get your beast off that little girl!"

"Bad dog!" the three-year old screamed valiantly, using her teddy bear as a shield. Aduna slammed the trunk and ran to her. Before he could get between them, the dog leaped and took Dawn by the throat! It dragged her along the pavement at an incredible speed. Goose pimples rose upon Gyasi's entire body at the sound of her horrific screams! Dianne was petrified! It would not take much for the dog to snap Dawn's neck! Dianne ordered him to save her, which fortified his resolve to challenge the animal bare handed! Then the woman dove, grabbing the angry beast and wrestling it to the ground. It calmed down without a fuss, and she held it by the leash, waiting for the ginger head fellow.

Dawn rose to her feet looking for her teddy bear. She seemed calm, as if nothing had happened. "Get my baby Aduna!" Dianne screamed hysterically, although the danger had passed.

Drake Shepherd took hold of the dog, shaking his head in mocked confusion. Dianne approached him while his partner watched nervously, apologizing for what she did not do or endorse.

Gyasi inspected Dawn. "I'm alright Daddy," the kid certified, trying to be helpful.

"You sure?"

"Yes. I am fine!"

"You look good on the outside," Aduna joked, "Now get inside the car!"

"Alright Daddy! Can you get the teddy bear for me please?"

Dianne argued with Drake, who was in forgetfulness. "You don't recall what you did? Spare me the drama!"

Gyasi saw the teddy bear lying snug on the pavement. He picked it up and took it to Dawn in the car. "Here it is!" he quipped. "It is still new and it is still mine!"

A supervisor, officials and crew rushed out for damage control. One man greeted Dianne, stepping off to the side to distract her. The supervisor took the animal and questioned its handler, "What happened Shepherd?"

The youth looked down and shook his head. "I don't know Mrs. Maynard."

She transferred her attention to the woman, "Lopez?"

"Well, don't ask me! I didn't just plain go loco!"

The superior shook her head understandingly and offered, "You did well today Lopez. I must commend you! You were brave, and we appreciate it very much!"

Gyasi waited patiently for the situation to get to some degree of sanity. Dianne ignored the official and went back to the handler. "Who was the person you spoke with before the dog attacked? The man who pointed to my daughter!"

The youth rolled his eyes to recall. "I don't know him. He doesn't work with us."

"He doesn't work with you? Why was he in your company uniform?"

"I can't remember ma'am!"

The supervisor butted in, shocked, "There was someone strange in our uniform?"

"An albino!" Dianne confirmed, "This man claims he doesn't work with you!"

"No albino works here!" Maynard replied anxiously, "Which way did he go?"

Dianne pointed in the direction and a couple men dashed forward, in a futile hope that he would wait for them to catch him! She smiled at the idea.

"Find that mystery man – Now!" the supervisor emphasized. "He didn't just disappear! Lopez, I'll call Horne! Go see what he has on camera! We need a face for him!"

One man turned to the youngster. "What did he say to you Shepherd?"

"I don't remember much. He told me to watch that kid."

"Are you playing the ass?"

"No sir!"

"But you can't remember anything! Did he hypnotize you?"

"I don't know if that's possible sir."

"Watch that kid? That's all?"

"Yes sir!"

"That's it?"

"That's it sir."

"And you think 'watch that kid' means to set a dog on her to break her neck? You're gonna wake up behind bars tomorrow kid! You know that?"

"I figure sir."

....

The next day...

Aduna Gyasi took the prototype off his desk and spun it around in his palms, observing from every angle. He smiled at his own genius. Dianne came with the cordless extension. "It's yours," she said, a strained look on her face.

"Who is it?" he asked puzzled. No one ever called him on the house line! Dianne left him to find out for himself. He put the receiver to his ear. "Hello?"

"Hello! Dr. Aduna Gyasi?" a chirpy voice replied from the other end. "I've been trying to connect you for a while!"

"You have?"

Before he could ask the pertinent question, the jolly fellow continued, "James Carter told me about your works! He gave me this number. A fine coincident it was to hear Dianne Scarlet pickup too! I know that voice!" He paused. Gyasi was about to speak, but he started again, "She's my old school mate you know!"

"Is that so?"

"Sure! We have a history together," he said matter-of-factually.

"I see."

"I suppose she's a Gyasi now!"

"Sure she is."

The caller paused and Aduna took the opportunity to ask, "Who are you though?"

"Oops! I forgot my manners! My name is Tutu Babe and I'm an entrepreneur!"

"You *are* Tutu Babe?"

"'The people's billionaire'! In person!"

The wizard grinned, "I am guessing that we have business to talk."

"We do Mr. Gyasi! How soon can you come to my office?"

"As soon as you tell me where!"

"Great!"

Gyasi hung up and hurried to the living room, looking for Dianne. She was not there and he muttered to himself, "Where the devil did this woman go?"

She came up behind him. "Where did who go Aduna?"

"*You!*" he tested.

"I go to everywhere and I come from everywhere in this house doing stuff! You go to your office to work, to the kitchen to eat and then up the stairs to sleep!"

"I told you we could hire a maid!"

"And I told you it would compromise our project! That was not even the issue here. The issue was simply me answering the question of my comings and goings!"

Aduna cackled, "You are spirited!"

"That's why you married me – right?"

He shrugged, "Could be! And I have news to fire up your spirit some more!"

"I already guessed it!" she returned in a calculative way. "That Tutu Babe is a back-biter and cheat!"

"Why do you say that? Did something happen between you two?"

"No way! Side-winding serpents don't fool me!"

"Are you telling me not to do business with him?"

"*No*! The man has resources! Just be careful! When you approach it from the front, watch the tail behind a scorpion!"

   In less than two hours, he sat before the entrepreneur, in a spacious office that looked like a club. Gyasi did his best not to chuckle when he saw the Scorpio sign on the wall behind the big man. Dianne warned him about 'the tail at the back'!

Tutu Babe showed interest in everything the little man presented to him. Aduna Gyasi accepted each offer without trying to negotiate. He never asked for too much. Tutu Babe liked that about him!

"I can tell you're bound to do very well Mr. Gyasi!" Babe surmised. The wizard thought about his statement and replied, "Financially? I can make useful products. My priority is not extreme wealth."

"What could be greater than making a billion dollars?"

"The ability to live well without money," Aduna countered knowingly. He decoded the underlying reason for Tutu's suggestion of wealth. Dianne warned him and that was his due diligence!

The big man frowned, "That impossible remark confuses me! However, you and I will do good business together. I don't only want to buy rights to market your inventions. For what Carter told me about your genius, I want you to work for my company! Create your incredible products and I will do the rest!"

"I would not consider working for anyone Mr. Babe."

"You really mean that?"

"I will not be programmed."

"Someone must take control to keep order, Aduna Gyasi."

"In my orbit, I keep the order."

Babe stopped to think, trying to figure out this little man. "You know how much you could make?"

"I do not. It is not relevant."

"I see," Babe nodded understandingly. "Your priority is not about money. I have fame and I know what it takes to market anything! As long as I endorse it, or it goes through my cooperation, it succeeds! Money or not, you can't do better than that!"

Both men fell silent for a while. Aduna felt tension in the room. The big man was disappointed. Then he had an idea! He dragged his heavy feet from off his desk and leaned forward. "What about contracts? Would you consider that?"

Aduna paused, thinking about Dianne's words. He saw no better mutual benefit than to concede. "I would consider that Mr. Babe."

"Fine!" the big man said excitedly, reaching for the handshake already, "Then we're in business! I have many propositions to make and I have billions to fund the deals! You will create the greatest wonders for Tutu Babe Cooperation! I will begin the process of settling cost with you – although you don't care for money! Then I will move full speed to roll out these inventions as fast as possible! In the twinkle of an eye, you shall become the most popular household name on earth! So, now that you work for me…"

"I don't work for you."

Babe hesitated and then grinned sheepishly, "I know that Aduna! It's just a way of speaking! We haven't even began to sign contracts! We will in due course!" The big man sat grinning at Gyasi, watching him as if trying to see inside his head. "Carter told me that you do impossible things!"

"I would not agree. The evidence is clear. They could not have been impossible."

Babe chuckled, "I agree! People get excited about everything! Tell me something that Carter thinks is impossible!"

"He thinks it is impossible to program genes by training them without using synthetic means."

Tutu frowned, "What do you mean – or what are you implying?"

"We could create a human population of all-powerful demigods."

"*Damn*! You talk about programming genes, and propositions flood my head!"

"That part of my work is not for sale!"

"Seriously?"

"Seriously Mr. Babe!"

"Why Mr. Gyasi?"

"It is dangerous knowledge. I keep it in the bosom!"

"I hear you!" Babe accepted, watching the little man and trying to disguise the hint of malice in his voice. They went silent for a few seconds again, before Babe changed the subject. "Carter told me about that incident at Logan's yesterday!"

"Oh *that*? It was very disturbing…but happily no harm came to my daughter."

"Sure! It's hard to see any harm done to our children – I mean, I don't have any kids yet…but, you know what I mean." He clucked his tongue sympathetically. "Oh Mr. Gyasi! Watch that kid! You ought to watch that kid!"

Aduna froze! That three-word sentence reverberated in his head to the point where it felt like a migraine! *'Watch that kid'* was all he heard at the airport the day before! Then he remembered the watcher's warning; *'You will know by the sound of the echo'.* That was the echo! He was warned!

Aduna's bowels moved in him and he stiffened. Cold beads of sweat formed on his brow. He was sure that he would faint. He heard Tutu Babe calling him, "Mr. Gyasi? Are you OK?"

The little man shook his head and sat up straight. "Yes. I am just exhausted I think."

"I understand!" the big man replied eagerly. Then he rose to his feet offering a final handshake, "Sorry for keeping you so long!"

"That is fine Mr. Babe. It was a pleasure to come and talk with you!"

"And you'll enjoy working here, especially with my resources. You'll see!"

As Aduna turned to go, the swarthy entrepreneur called to his back. "Mr. Gyasi!"

The little man stopped and half-turned to see the scorpion behind him.

"I have to leave this thought with you. The fact that you offer something doesn't mean the people will buy. You should sell them what they want!"

"I know!" the wizard replied soberly. "What they want is what we teach them. They will not want what they do not know."

As the little man hurried to his car, he wondered if the same thing that happened to that dog handler happened to Tutu Babe just then. Maybe the rogue Watchers controlled that serpent too! How would he protect his family? His two daughters were human!

3

The world took notice! Aduna Gyasi was the sage of the age! With Tutu Babe Cooperation behind him, his products were extremely popular. He signed off on a couple contracts for the billionaire and those projects were going well too.

Dianne's words at the death of Lisa opened new trends of thought in the wizard's brain. The intuitive woman directed the god-man to the invisible mind of the universe. Her ideas transcended cultivated notions of divinity. It inspired him to transform and advance the old theory of God and religion. By the wisdom of Aduna, they would evolve from superstitious presumptions to provable and potent life principles.

His personal desire to decipher the mysteries of the world in which he lived motivated his philosophies about life. He shared them because of his passion, which did not always amount to good results for him! Many students gravitated to it. This infuriated Christian Churches and zealots who challenged the little creator head on! They urged him to desist, claiming that he was corrupting the minds of the people and condemning them to hell. For Aduna Gyasi, to know was the greatest thing. He refused to give in to believing sentiments. The wizard became the most controversial human being in recorded history!

The real controversy began when Mavis Stern, a North Carolina woman, called Reverend Beacon's television program at their discussion time. Hartwell Beacon was one of the most celebrated and educated evangelists in the nineties. He started The Church of the Risen Savior from Somerville Massachusetts. There he did his live evening broadcast every Sunday, in front of a massive gathering.

That first Sunday in June 1997, Mavis called imploring him to invite Aduna Gyasi to his program for a debate. At first, the knowledgeable minister declined but Mavis Stern found phenomenal support from the other callers plus the entire congregation!

Finally, the church would rise as Christian Soldiers to protect their children from demonic doctrines! Within days of the famous phone call, they formed a faction named Faith Defenders. This group sought to pit the greatest theological minds against the best scientists to prove God. Eventually, the great minister succumbed to the pressure. They contacted Aduna Gyasi. The wizard jumped at the opportunity to explain his ideas to the masses! The media caught the buzz!

That second Sunday, twenty four thousand gathered for the occasion! Television rights were sold out! At the introductory stage, Gyasi was laying the foundation for his argument. *"In recent times, Mankind raised the topic of Intelligent Design. Conventional scholars shot it down, but with taboo instead of reason. I believe that these principles were not all flawed. Understand that there is not always wrong or right! There are always dynamics, dimensions, perspectives and circumstances. Truth depends on the value of these variables. Let us say that 'God' is in the center. Around that center, there are East, West, North, south – and spaces between them. It is not wise to say that the way to God for all men is to travel west, just because you live in the east! When men worshipped the god Yahweh, thousands of years ago in the east, other people lived in Europe too! These people had to adhere to authentic conceptions of divinity. That was their truth. There is no truth to say that red is better than green, but there is a right to choose. It is an opinion, a perspective or a perception. It will be unjust to arrogate on it!"*

Reverend Beacon interrupted, just to 'make something clear'. "Dr. Gyasi! When you raise points, verify them with reference! The holy scriptures verifies all that we claim!"

"Do you have a verified bible now?" Aduna questioned sarcastically. *"Since we all presume so much, let the truth come out in the individual's heart! What can we verify? What do we see when we look around? Do we see God, or do we see Perpetual Motion? Do we see equations or do we see a judge? Is Perpetual Motion an equal cycle? What is justice and how does it perpetuate? I find that conscience is the one perfect attribute in humans. Love is the ideal culture. Conscience is love's essence and the three virtues of love are goodness, truth and beauty. Of these three, goodness is most important. I propose that the laws of the universes and their perpetual motions stem from human conscience or some higher vein of similar design!"*

"I say these things, not to disagree with any church, but if the argument disagrees then so be it. I do not believe that your doctrines and rituals will get me to 'Heaven'. I propose this new truth as the beginning of entry to a higher realm. My words disagree with many scientists too! Conventional sciences, from a visible universe perspective, cannot lead to the absolute human advancement. I do not believe that a war with some space devils is imminent to decide the future threshold. The universal mind, unseen and eternal, creates the destiny of all. My faculty as a science wizard is God's signature confirming my purpose!"

Beacon chortled sarcastically, "And what is that purpose Dr. Gyasi?"

"My destiny is to remodel the human prototype!" Aduna replied easily. It was not arrogance. It was not his attitude! It was his blasphemous words!

"What?" Beacon asked, stunned, "Are you God now?"

Before the wizard responded, Beacon added, "I cannot do this! I am in the presence of Satan and I'm going to quit!"

The chubby reverend exited the stage, but his actions incited the masses. They reciprocated with violent rage! The first missiles came...plastic bottles, food containers, food...any solid object available! Every soul in harm's way scampered for cover. Aduna was the last to react! As he turned to exit, a chunk of brick crashed into the side of his head! He staggered off in a daze. Hot blood gushed upon his right shoulder blade. Another hard missile hit against his ribcage, piercing his jacket.

Two security workers grabbed his arms, directing him to his car. They saw a dozen zealots coming for the wizard, and it caused them to panic! They ran! Aduna was in trouble! The sea of fury covered him, raining enthusiastic blows that some hit their cronies accidentally. Soon help arrived to pull away the punishers. They pulled the wizard from the bottom of the pile. He arose, battered and bruised...but alive!

Gyasi dusted himself off and headed for his car. There came another wave of bodies to him. This time it was the media, come to corrupt and spread the twisted news!

....

On Monday afternoon, he awoke, struggling to open that one good right eye and forgetting what happened to the other. Gradually the memories began to flow. He was alive!

"What time is it?" the little man asked himself. His default voice responded, *"The time is 4:28 P.M."*

"Huh?" he reiterated in shock, rising up on his elbow. The day was already done, and he did not do a thing!

He heard the television playing and realized that he was not alone inside the room. Aduna looked to the corner. Dianne sat in front of the big screen. "What are you watching honey?" he asked childishly.

"*You* honey!" she drawled back with irony. "But I can't see you now…there are too many people on top!"

Aduna blinked his good eye clear to make out the images on the screen. It was a news recap. They were pulverizing him, and Dianne winced at every blow.

"Golly!" he mused in a lame way, "It did not look so bad!"

"Well you're the scientist, always not seeing what we can! It's hard to believe you survived!"

"Well I did!"

"Uh-huh! Is that how they congratulate you for winning the debate?"

"Very much so! See, I am a debater, not a fighter! I did not see a bloody crusade on the horizon!"

"The Jamaicans say 'trouble doesn't set up like rain'! You wouldn't see it coming! Next time, before you go to the inquisition with pagan philosophies, remember what happened to Jesus! He was the first guy who tried!"

Suddenly a thought came to him and he asked in a very urgent tone, "How on earth did I get home?"

"You drove Aduna! You even rang the doorbell before you passed out!"

"I cannot recall a thing!"

"I *know*! After taking that scary red brick to the head, who would? I'm flattered you remembered me! *'Honey! Honey! Come quickly!'* were your words exactly! I bet you told everyone over there *'I am alright! I am alright'*, desperate to not go to the hospital!"

"Maybe I did too!"

"Good for you! You'd be swimming in poisons by now!"

"What a scary thought! Profanity!"

"Carter came to patch you up! He put a shot on you, as I requested. Something to let you sleep longer!"

"Oh…I see."

Dianne rose from in front of the television and went to look through the window. He waited, knowing that she had 'stuff' to say. "Tutu Babe called me today."

"*Oh*? What about?"

"*You* Aduna! He wants me to talk sense into your head."

"He proposed to use my Code666 to help the élite enslave the masses!"

"You know I'm with you on that point! He said something else that I agree with."

"What is that?"

"Stop trying to teach unconventional stuff. You know what you know Aduna. God gave you eyes to see. The masses haven't ascended to that degree. Divine light shines on you but it doesn't mean you're responsible! Most humans are irredeemable. You cannot save them. Let them save themselves! Teach only those that come to you. Concentrate on the Cyber Baby Project. It's all we have to do! Don't draw attention to us! The zealots that creamed you were no Christians! Someone sent them to cause mayhem!"

"Maybe you are right. Wolves are among the sheep. Considering the pains that it could save me, toning down is a good idea!"

Dianne giggled unwittingly. "It'll save pains, won't it?"

He got the joke and afforded a snicker too. Then he admitted soberly, "Sometimes I wonder if I should scrap the project and live a normal life!"

"There's no choice! It's what we are! We're not normal!"

"You do not understand what might be at stake!"

"Whatever is at stake will not make us what we're not! We're different! Nature is our coworker. It teaches us that once the order is set there's no changing it. If your roof is broken, it doesn't mean the rain should stop. Whatever we have at stake is like that leaky roof. Fix it Aduna!"

4

He was finalizing the super anti car theft system for the cooperation when Babe came and took him by the shoulder. "Aduna! There's something you gotta see in my office!"

Carter was talking to the god man when Babe arrived. He frowned, watching them leave. "Come on Carter. You might wanna see it too!"

The television was on in Babes office. "Sit down, and watch this news report! Denzel recorded and sent it over for me. We had a bad egg in our system gentlemen! It was leaking to the media!"

Babe started the tape: *"Here is something you won't see every day! You won't see it anywhere but on this network. The little scientist whom we know as 'the god man' has outdone himself again! He stunned the science world in a demonstration that was supposed to be 'hush-hush'. Of course we have our sources to get you the information!"*

"The unconventional Aduna Gyasi has been doing the impossible since coming onto the scene this year. We heard the rumors about his superman code triple six, although no one is taking it seriously! What we are about to show you will blow your mind though! It is tangible and provable. Therefore, it happened!"

"Weeks ago, we told you about some speculations in the realm of science. His confused counterparts believe that Aduna can determine the outcomes of seemingly random events. Some say that he can write holographic scripts to alter reality. He calls this principle 'Reality bending', but the flabbergasted fraternity calls it 'Science Magic'.

"As we speak, a tiny islet near Tierra is missing! Tierra is one of many uninhabited rocky islands in the region of Sub-Saharan Africa, which Spain owns. Moroccans use these islets to cross over illegally into the European nation. Rumor has it that the great scientist performed a famous, 'cut and paste', which the authorities were trying to keep secret! We do not know if Aduna did perform this feat for sure, but what we know is that the islet is not there anymore! Watch this clip! There you'll see the images for yourself! Picture 'A' is a shot of the location before the islet disappeared. Picture 'B' is a live shot. No islet – See? Looks crazy? It gets crazier!"

"In Jamaica, way over in the Caribbean, they found a new islet near the fisherman's Pedro Keys! The Pedro Keys anglers woke up one morning and there it was! Is it the missing islet from next to Tierra? Well, it has the same dimensions in terms of size and shape, but most importantly, it has vegetations that do not grow in or around the Jamaican region but are identical to those that grow in and around Tierra. So, is it the islet then? You decide for yourself. Ann-Marie Facey, WOW News!"

"Well," Carter quipped sarcastically, "That's as secret as it bloody gets!"

Aduna shook his head disappointedly, "With due respect Mr. Babe! I do not care who hears, but I swore to secrecy for a given period. You need to know the people you hire!"

Babe turned his bloodshot eyes to the little man, piercing his skull. It seemed almost like a challenge, but finally the big entrepreneur admitted, "You're right Aduna Gyasi! You're so doggone right! The only person we could not ask about this leak is Thornton. Did you hear what happened to him?"

"Thornton? No! What happened to him?"

"I heard he died of an overdose this morning!"

"Oh! That is sad!"

"It is, but it's done already. This morning the police found the tapes in his room when they went to investigate. He was the leak!"

"He was?"

"Yes…but let's talk about how you're doing!"

"Everything is going well Mr. Babe! That means I am well too!"

"Good!" the big man said, sounding like he had more to say. Gyasi watched and waited. Carter excused himself and left the room. Tutu finally obliged. "This is another attempt to have you pull your cyber gene technology from the closet and put it to some good use. Use it instead for something that's coming up, which the US Government will contract us to do!"

"What is that?"

"I can't say yet, but we're going to need your input!"

"Fine! If it is worthy, I will assist with my technologies."

"Wonderful! Those are the most encouraging words I ever heard from you! Besides, any project you undertake in secret, with information you're hiding, will make you a rogue scientist in the circle! That's how it works. I'm sure it will affect you and your family in a negative way."

"Will it?"

Tutu Babe did not answer him. He knew the answer was obvious! "Another thing Aduna! Since you came to offer your expertise, you've excited me beyond my wildest dream! You've excited the entire fraternity! We feel as good as those of any advanced dimension because of you! However, some of what you have demonstrated is beyond our time. We cannot tell all! We must give the lousy populace time to catch up, although it means waiting to make more money for me! That sucks!"

"That is an honorable decision Mr. Babe. I agree with you."

"Fine! You must sacrifice when your time comes to do so too! The circle of wizards has agreed. We even fixed a date for it! We ask you, do not officially unveil 'Science-Magic' until June 10, 2016, which is almost two decades from today. Even then, you shall reveal it only as esoteric wisdom, entrusted to a selected few. The initiates will receive it as well as they can. I only hope that by that time other students will be able to decipher it! You cannot always be the only one on earth Aduna! I don't mean it in a spiteful way!"

"I see your point Mr. Babe. It is a reasonable one!"

When the little man left the office, he found Carter waiting at the door. "Aduna!"

"Huh?"

"I know he was badgering you about triple six!"

"He always does!"

"Do you know how much money you can make working with him on that?"

"I have no idea Carter, but it does not matter to me!"

"You're bloody crazy! Take it from a friend!"

"I hear you."

"Here's a suggestion! Since we won't work with Babe, use the cyber codes to strengthen your posterity and mine! It can't be free-for-all! Imagine our children as gods on earth!"

"Carter, I discovered something wonderful and if I would use it, I would do something wonderful for humankind as a whole!"

"Which would be a challenge to the earth and the heavens! You would regret it!"

"Why would I regret it Carter? Would you tell on me?"

"What kind of question is that? Do you think I would betray you? I'm saying it would be dangerous! It is your code. You have the say, but Babe would do anything to stop it if you tried. The Watchers and extragalactic beings…they know everything!"

"Yes, but they cannot tell you everything. Moreover, it is not there say!"

"Not there say? What difference does that make? There are rogues among them, and those have their own agenda concerning the bloody earth!"

"They have their agenda and I have mine! My agenda is to free humanity!"

"Can you fight them and win Aduna? They might not be able to lay hands on you directly. What if they share their technologies with bloody humans to fight you with?"

"Your *what if* is a good question…but it seems I have no choice. The project is my destiny! I live by it or die by it!"

"What about your wife and children?"

"Their destiny is to be Aduna Gyasi's wife and children!"

....

On June 21, Gyasi finalized the car security system. He was going to the office when Carter stepped into his path, "Slow down partner! Why are you moving so fast?"

"I always move fast Carter! There are many things to do!"

"There's always much to do, but I believe in slowing down anyway!"

Gyasi chuckled, "Maybe, but I enjoy not slowing down. See?"

Carter shook his head, "There's no getting around you on points. Is there?"

"I only make the best points possible Carter."

The biologist changed the topic. "Aduna, you need to take that warning from the Watchers seriously!"

Those words stunned Aduna but he suppressed all show of emotion. "How did you know about that Carter?"

The lanky man frowned and then replied awkwardly, "What do you mean? Don't you remember telling me?"

"When? I cannot recall. I am sure I did not tell anyone."

"Don't be too sure!" the tall fellow warned. Then, after giving it a little thought, Carter added, "Maybe you were talking in your sleep then. You told me everything that day when I had to drug you."

Gyasi leaned his head sideways to think. "Did Dianne hear me talk about it?"

"No. You didn't want to tell anyone else. Remember now?"

Aduna paused to think again. "I do not remember."

Then his sixth sense told him, *"Dianne is on the line!"*

"OK." The call came through.

"Aduna! Why did you pick up Dawn without telling me?" Dianne agitated.

The little man was baffled. "Me?" he asked awkwardly, "Pick up Dawn?"

"Didn't you hear me Aduna?" Dianne shot back impatiently.

"Dianne!" he replied, feeling his heart begin to throb, "I did not go to that day care! Someone else must have…"

"But they're sure it was you! They…"

"Let us not argue!" Aduna advised soberly, cutting her off. "It was not me!" he scratched his head and glanced around as if to discover his own thoughts. Then he asked, "Where is Ray and Charm?"

"They're here! Oh my goodness! I'm calling the police!"

"I am on my way over!"

Carter watched his friend intensely. As soon as the wizard finished talking he asked, "What's wrong?"

"I have to go! Someone looking like me abducted Dawn!"

"Bloody hell! I'm coming with you!"

....

They pulled the back doors on both sides of the green Ford Explorer and stood observing her. She watched in terror as they removed their faces! The little one who had been her father transformed to some cock-eyed devil she never knew. When the tall one took off his Carter James face, she knew whom it was! It was that dog-handler from the airport!

"Bad man! Why are you not in jail?" the child asked dumb-founded.

The lanky man hyena cackled, "Jails are not for connected people!" then in the same breath he told his partner, "We'll take her to the basement! She has a sacrifice to make!"

"Let me go you big monster!" the child screamed in a demanding tone. The tall fellow took a glance at her and hissed disgustedly. "I won't tolerate this noisy little brat, 'cause I'm not in the mood!"

He lifted his right trouser leg and unsheathed a huge dagger from his boot. Drake patted her cheek with the side of the blade that she shivered uncontrollably. "One more word out of you, and I'm gonna slice your face off!" he informed her curtly.

She bit her lips to make sure her words did not escape. Floyd chuckled and the tall fellow turned a malicious eye to him. "What's so funny?"

Floyd was about to point to Dawn but then he changed his mind and said, "Nothing!"

Drake Shepherd grabbed her legs, which Floyd had strapped together with duck tape, and pulled her from the car. "I said to strap her hands and feet! I don't see why you had to strap her all up like a fuckin' zombie!" he agitated. He hoisted her over his shoulder, while she wriggled her body, trying to resist. The short fellow watched, amused.

As he hustled to the house, the front door flew open. A woman stood to the side to usher them in. Dawn realized where she was! "Where is Dr. Carter?" she demanded.

The woman ignored Dawn's outburst. "There's an office in the basement where the doctor works. He sent someone to do the transfusion!"

Drake Shepherd headed down the stairs with the little man tailing him. Dawn held her head up, scolding Floyd with her eyes. Floyd ignored her. "What transfusion is she talking about Drake?"

"Carter's kid was born with some rare disease called PKD. She needs a transfusion every few weeks. Problem is that her blood type is B negative!"

"So what if it's B negative?"

"Means it's harder to come by Floyd! There're not enough donors in the family."

"But her dad can donate for her – Right?"

"It's kinda technical! He has to keep his own in reserve for some possible bone marrow thing! I don't know!"

"Well how do we know that this kid's blood type is the same as his daughter's?"

"We wouldn't fool! Dr. Carter would! It's not even the same. It's O negative!"

"That's confusing! If she's O negative, what are we doing with her?"

"You ask too many questions! I don't give a shit if it's O negative…A, B, C or fucking Z minus! Do I look like the doctor to you?"

"No," Floyd replied, embarrassed. His eyes caught Dawn's and she stuck out her tongue to mock him. "Stop *that*!" he snapped distastefully.

They entered the office space. A man in white walked over to greet them. "Great! You're here! Just put her on the bed over there! The organs belong to me! That was the agreement. Tomorrow you boys come and pick up the rest for closure!"

....

Dianne sat in the back seat, tearing at her hair. Carter watched her through the rear-view mirror. He turned to catch Gyasi's eyes. The god man held out his palm to indicate that he should leave her alone. Carter nodded understandingly and switched on the ignition.

"I do not know where else to look Carter!" Gyasi said in exasperation. "Have we not searched everywhere? The police are combing the entire city! Where could she go?"

"I don't know! It's bloody hell! They framed you Aduna! People will believe you did it for as long as you live!" Carter flashed an eye at Dianne, and then pretended not to notice her. "I wonder..." he began hesitatingly.

"You wonder what?" Gyasi quizzed, pressuring him to speak.

"I wonder if 'those entities' are behind it..."

"Who?"

"The ones that threatened you..." he said feigning a whisper, "I mean the Watchers!"

"What Watchers?" Dianne chipped in bewildered, "Who got threatened?"

Carter eyed Aduna apologetically, just so he could see that it was an accident. The god man strained his neck to see Dianne. "I had a strange visitation recently. It came from some unearthly source…"

"The Watchers you mean?"

"Yes. They feel threatened about my superman code."

"So they threatened you?"

"They threatened my family!"

"What the heck? Why didn't you tell me?" she questioned with much less venom than Carter expected. Then she began to tear at her hair again. Carter watched her reaction and frowned. Their level of trust was peculiar to him! He thought for a while and then said testily, "If the Watchers came, it is for a logical reason! They know much more than we do! This is why I advised Aduna to scrap whatever he's doing!"

"The project remains!" Dianne interjected stubbornly, shocking Carter. Then she reiterated with unshakeable resolve. "We will not scrap a thing!"

Dianne said too much, but she could not take her revealing words back, so they waited for Carter to ask, "What project are we talking about? Do you have a cyber genetic thing going on without me in it Aduna?"

"All this talk is fatiguing Carter! Right now we need to find Dawn!"

"They better not harm my daughter!" Dianne hissed matter-of-factually.

………………………

Next day, Aduna did not go to the office. The police came to question him a second time. Then they decided to arrest him! The day care workers insisted that he came to pick up Dawn, and that he drove the green Ford Explorer.

When Dianne realized the intention of the officers, she had no choice! She took up the phone and called Tutu Babe! Then she jumped in her Mini Coop and followed them down the three hundred meter driveway.

Before they reached the gate, the police car stopped in front of her and Aduna jumped out, scrambling to see something that someone dumped next to the fence. Instinctively, she stopped her own car and went after her husband.

Aduna stooped over Dawn's still frame. One of the officers walked up behind him. "Jesus Christ!" he lamented, "It was not here when we came in!"

The other cop walked across and chipped nonchalantly, "Seemed like they cleaned out her inside! There's not a drop of blood in her!"

Even while the god man crouched over in unbearable pain, the insensitive officer was poking the grieving father, "Now, will you tell us what you did to this little girl?"

Dianne was scurrying past him to get to Aduna. She paused by the cop, pulled back her left arm and let fly! Her bone cracked on impact, as her fist connected with the officer's mouth. He fell flat on his back holding on to his face. *"What?"* his colleague asked in astonishment. Gyasi did not realize what was happening. Dianne went to his side, oblivious to the pain of her own cracked knucklebone. Then they held on to each other, crying and begging the dead to rise!

The insensitive man rose, spitting blood. His colleague bit his lips to contain a snicker. "Are you alright, partner?"

The victim looked at his attacker spitefully and then at his partner, "Yuh! Dat meanz zhe's yoing to zail too!" he informed, his battered lips making it hard for him to pronounce the words.

His phone rang and he fumbled for it. "*Heyyo?*"

The other cop watched him, still fighting hard not to laugh. He never saw a mouth swell so big so fast in his life!

"Oh! Mr. Babe! How you yoing zir?"

When the wounded officer put down the phone, he turned to his partner. "Tome on! yet's yeave 'em here! HQ's zending in zome other people!"

"Are we going to make the arrest?"

"What ayyest? We vind the twiminal who did this!"

"You sure? Mike Tyson's sister over there did assault you!"

"Yeave her ayone! Zhe's drieving!"

In the spirit realm, the god man wrought marvels by way of Mind-Matter Principles. The scientific community kept them hidden, while conspiring to drain his brain for the wisdom of Aduna Gyasi. They feared what awesome effect such knowledge would have on the shepherded populace. In the conventional sciences, the god-creator made super high-end Servant Robots for billionaires.

He spent much time teaching and holding press conferences to share the light. His lifelong friend Carter James became an avid student of Thought Science, which seemed beyond him to grasp. Carter introduced an African student named Samuel Uche to the team. Uche was a very smart youngster and the god-creator took a liking to him. The youth had a passion for the study of sound frequencies and radiation. Gyasi became a mentor to Uche, educating him in the field of his passion.

Although they kept Thought Science hidden from the people, high profile scientists and researchers worldwide wanted a part of it. Tutu Babe organized the famous Boston Science Conference. Aduna Gyasi was the speaker that everyone went to see and hear! As much as they learned, it seemed inadequate. Either they were unable to change their preconditioned mindsets or Aduna was deliberately teaching them secrets of lesser value.

The God-Creator stepped to the podium, dressed in 'neutral-cloth', which was not yet introduced to the masses. This material kept the body at the perfect temperature regardless of the surrounding conditions. Aduna went determined to condition minds and win support for his *ungodly* way of thinking.

"What is truth? Should we cast our beliefs in concrete? Truth is nothing but conceptions that lead to

higher ones on to all knowing. Thinking Thomas would say, 'The same things appear different to different minds and even to the same mind at different times'. The threat of stagnancy is real to those that stop at a given truth. When your truth comes into question, it is no longer relevant. You must advance to superior ideals and comforts. That is what it means to advance. That is the definition of a 'heaven'. Continue to conceive and stop 'believing'. For sentimental fables, we deny what we can prove, saying 'blessed are the believers'. I say believers are sheep dependent on shepherds. A shepherd's best friend is the butcher! That is obvious!"

"By redundant conceptions of God and Satan, we confuse ourselves. We diminish our capacity to grasp natural laws. We will stay at an elementary level, until our truth advances. Many of the things I know I cannot teach!"

"Fate and destiny reside in the universal conscience of perpetual motion. They operate in the order of cause and effect, to the equations of cosmic cycles through the ages. There is no way to succeed and no heaven or hell outside of universal law and order. Internal and inherent justice attains in all things. In the human realm, intelligence, not ignorance, must drive us to heaven's threshold. Expel all demonic thoughts!"

"A demon is the same as any cult. It is an adverse idea or a legion of thoughts. A mere virus it is, just as on the computer. You must click to give it permission to affect you. You can reduce to mediocrity and servitude. Evil thoughts can possess you, if you 'click' into them! Viruses and sorcery are all around you, deliberately placed as mass gins in your way. Knowledge of the unseen makes a man into a light being. A king will become a god, determining human fate and destiny."

"Do onto others as you would have done to you. That is a universal Law of conscience, which you conceived by yourself. It is proof of God Internal. Yet, you have clicked into many gins that move you to defy this truth. It is obvious in religions."

"We speak about God but who or what is God in perspective? Look at the physical world around you! In this visible realm, Omnipresent God can be no less than all in all. In the invisible realm of thought and mind, that drives us, God is in the conscience. Conscience qualifies humankind for higher intelligence and authority to rule the sphere. Intelligence and authority in the unconscionable mind is anarchy and anarchy is satanic. Human conscience will create and order destiny in our universe of perpetual motion."

"We were not born to any religion. We were conceived as humankind. Nothing outside of our humanity can condemn us to hell or not. Our humanity comes with the internal compass to our heaven. It is our conscience! Nothing on earth will transcend the conscientious authority of Man."

"This Mind-Matter Science is not to bore or burden humankind. It is to show you how the cosmos maintain perfect balance. We must learn peace with the order on our plane. The Law of Love and the Universal Conscience is our path to Heaven."

Aduna Gyasi paused to catch his breath. The room was silent. The scientists looked on, wondering what to make of his diatribe! He performed wonderful things that they were desperate to learn. If some strange astrological belief were the answer, they would be interested! The more they listened, the more they discerned science between the follies of his fables. Then, from the back of the room, someone shouted, *"Dumb stuff!"*

A few people began to jeer. A wiry man put his hand up. Aduna pointed to him. "Dr. Gyasi!" he asserted matter-of-factually, "I came here expecting to learn about science, but what this seems like to me is speculative pseudo-religious conjures!"

"I know!" Gyasi replied sarcastically. "What else could it sound like to you? Things will appear to be magical until the unenlightened realize the natural reality in them! That is, if it is within his ability to conceive."

"That's feisty!" the man accused with hostility.

"How did you get inside here?" Aduna asked befuddled. "This conference is exclusive!"

Then the missiles came! Zealots were there for the crucifixion! Running would not befit his style, but with no better thing to do, he charged from the limelight and headed back stage. There he sat with his palms holding up his chin to think. This time he took a good look! The troublemakers were just a handful of people, deliberately scattered through the crowd. Most likely, they were not even scientists! Someone was working with the Watchers to do these things!

He heard an official onstage trying to regain order. That would be easy. His detractors had only sent the attackers to make Aduna look bad!

Then the media came to *twist* the truth for mere drama!

Since Dawn's death, he had the media around his neck like an albatross. There was the accident at the airport, and then someone kidnapped and murdered her. The day care custodians were sticking to the initial story that seemed like an indictment against Aduna. He could not blame them if that was what they saw. He could blame the media for the rhetorical questions that only served to create adverse believers in the masses!

When they finished with him, he headed for his car, looking as dejected as he felt. Carter was there waiting. "Look at you! You seem like a man without a bloody nation! Don't you see that they're destroying you? Is it worth it mate?"

Aduna opened the car door and went inside. Carter leaned his head down to window level. "Your cyber-baby project is causing the problem! The Watchers don't like it and you know you cannot fight them!"

Aduna started up the engine.

"Oh bloody hell Aduna Gyasi! Dawn is dead! What else is gonna happen? We don't know…but is it worth it mate?"

The god man put the car in reverse.

"Well, tell me Aduna, are you going to scrap the useless project?"

"Absolutely not!" the god man snapped back and drove away.

"There will be bloody hell to pay my friend! You could have ended up in jail! Tutu Babe might not help you this time!"

....

"It's going to be a wonderful fourth of July this year!" Dianne exclaimed confidently, trying to distract Charm and bring a smile to her face.

.

"Why do you say that Mom?" the seven year old asked, looking through the window as if to find anything special about the day. Dianne peeked at her. Charm was still thinking about it! The little woman turned right on to the street at the bottom of Lechmere, pulled up to the big stop light and switched on her left indicator. "I say it because I'm optimistic. I want everything to be perfect!"

"What can ever be perfect Mom? Dawn is dead!"

"I know honey! I miss her too! Does it mean we should forget about making each other happy?"

Charm heard the question and it challenged her comfort zone. It was right and hurtful at the same time! How could she just switch off and go into happy mode?

"No Mom."

"I guessed not too! Since we're here together, let us do what Dawn would do!"

That last sentence seemed to work. Charm forced a smile, and a glint of light hit her retina. Dawn was always so cheerful, trying to make everyone laugh with her.

Dianne hit First Street and pulled up into the entrance to the mall. She grabbed her ticket from the machine and headed to the parking area. "So, are we going to have some fun?" she asked, looking around anxiously for where to park.

"I guess so," Charm offered defeated.

As they went through the first floor of the mall, a magician jumped out in front of the people ahead of them. He did tricks that fascinated Charm. Dianne looked down at her daughter's face and smiled, satisfied.

The funny looking clown character picked a flower from behind a young woman's ear. Then he cut a rope in two and made it mend miraculously. Charm concentrated hard, trying to find the source of his magic. She could not. It seemed like this man was real.

Then he stepped to her. "Good day little miss!"

"Good day sir!"

"Do you see my fingers? How many do you see on this hand?"

"Five!"

"Good! High five! Do you know why I'm always happy?"

"No!"

"Because I fell off a high place and survived. Now I'm so glad I'm alive that I celebrate everyday!"

"*Wow*! You do?"

"Sure I do…and I still love high places too. Do you?"

"No! I am afraid of heights!"

"Good! That's what I'm talking about! Be very happy, but be careful too!"

Charm shook her head, confused. "What sir?"

The clown snapped his fingers, "I said nothing! That's the trick!" Then he plucked a rose from behind her left ear and walked away. "That's the trick!"

Charm giggled, watching the fun person leave. She watched him until Dianne chipped in, "Let's go get your shopping done!"

They went in and out of different stores looking at the various fashions before buying anything. When they got to Aeropostale store, Charm was delighted. She loaded Dianne up and went to the check out with her.

The Hispanic woman peeped through her glasses inquiringly. "Good afternoon ma'am! Do you by any chance have one of our store cards?"

"Yes, she does!" Charm chipped in before her mother could answer.

The cashier smiled, "Thank you young miss! You've been a great help!"

Dianne smiled, "So, let us see the damage here!"

The woman finished checking and offered, "Three hundred and sixty ma'am."

"OK!" Dianne mused, surprised at how reasonable it was. Then the woman looked above the tip of her glasses conspicuously, "Lady! Where did the little girl go?"

Charm was gone! Dianne's heart jumped to her throat. She was about to go back into the store, but the kind-hearted woman suggested, "If she's in here she's safe!"

Dianne got the message. She made an about turn and headed outside. "I'll hold on to these for you!" the concerned woman offered sincerely.

As Dianne entered the walkway, she saw people running from all directions to a section of the rail.

Charm was standing on top of it, above a thirty feet drop to a hard floor! Dianne was about to shout but changed her mind. It would probably startle the seven year old! No one could get close to her, for she was threatening to jump! What was Charm doing?

Dianne parted the crowd to get through, "Excuse me! That's my daughter!"

When she got to the front Charm saw her and locked eyeballs. "Mom!" she asserted ecstatically, "I'm going to take my happy fall now!"

Dianna's heart leaped! There was only one thing to do! She charged forward but not before the little girl let herself fall. She reached forth desperately, trying to grab Charm's left angle. Too late! She could not hold on! Inadvertently, Dianne's fingers scratched Charm's leg and she felt the skin peel off into her fingernails. That was her last touch!

She watched her daughter fall backwards, as if in slow motion…smiling contentedly to the point of impact. Charm landed flat on her back, and blood spurted from her mouth and nostril. In the midst of desperate pain and fear, Dianne remembered the man who pulled the tricks! "That magician!" she bellowed in anger as she charged down the escalator to the first floor, where Charm lay still. "He killed my baby!"

....

After Charm's death, the media went wild! They never relented, raising questions of child abuse and neglect. Aduna Gyasi and his family found it almost impossible to stay beneath the radar. He thought about pulling the plug on the project. Carter James was happy with that idea, until Dianne shocked them again! "The project," she insisted, "isn't something we chose. It's what we are! We live or we die! We are the project!"

"But..." Carter besieged stunned, "The Watchers..."

"Fuck the rogue Watchers Carter James...and fuck anyone who works with them! Whose side are you on anyway?"

Both men turned to her in astonishment. They never heard curse words coming out of her before! Gyasi always knew that she was super strong. Now she was exceeding, by far, what he would have imagined. Their second child had died and their reputations destroyed. She was staying on course anyway!

"Dianne..." Aduna questioned speculatively. "Can we handle this situation?"

"Can you? Losing a child is the greatest pain! We already lost two! When your family is all dead, you will still be alive! They need you! Continue doing what you're here for Aduna! We will remember you in our next life!"

From that day on, no one questioned the project anymore. Carter agreed to work along with them and forget about 'his fears'. He suggested that they send Ray away to his grandparents in Bombay. That sounded like a good idea, and the Gyasi's hurried to get last minute tickets. The next day, Ray Gyasi left the United States.

He never made it to Bombay! Somewhere along the way, he disappeared! Carter seemed affected more than the Gyasi family. He hounded the police for results and chided them over their approach to the situation from the very beginning. Thanks to him, the lawmen finally saw the possibility that someone was trying to destroy the Gyasi family. That was after they already did!

Ray was the eldest, and now the last of their three children. They had nothing left to live for but the project! Aduna knew that his secrets were keeping him alive. What kept Dianne alive, he did not know!

The investigators were sure that Ray took the flight but they combed the city anyway, searching for the nine-year-old Gyasi.

From that day on, Aduna Gyasi stopped caring about what people thought! They went home that night feeling weary of life. Dianne watched Gyasi's face. She knew exactly what he was thinking and opted to ask the question for him. "Do you believe we did the right thing?"

"How can we be sure? We will find out when the time comes."

"That's what I'm thinking too, but I hate leaving anything to chance."

Aduna leaned his head sideways, in a meditative way. Then he said, "Ray is like me, but he has far more faculties. He is way ahead of his time! This is the way of perpetual motion! We have made great sacrifices. Is it to pay for Lisa's life? I do not know! All I know is that we have paid at a cost of two for one! No one knows the faculty in this kid but us. Ray is safe!"

"Yes. I know we have paid. I wish you had listened to me at first, when I asked you to plant the seed in me!"

"You would be dead Dianne!"

"And you would have Lisa to work with. Most of all, the children would be alive!"

"Maybe the only difference would be Lisa changing places with you. Everything else would be the same. Remember too, the girls were human. How could we save them? Let us stop speculating, rise at this point and defeat our foes with science!"

"And the Watchers that declare war on us?"

"I dare to think that I have the faculty to challenge them! I will be one step ahead, perpetually! I should have done the same for the girls as I did for Ray!"

Dianne breathed in deeply. "You were not sure about the code dear. We took one risk. It turned out OK, but not as that which is yet to come. One day our girls will remember this past life. They will reincarnate to the form we created. At that time, they shall understand the sacrifices we made."

"Ray understands."

"I know!"

A thought struck Aduna and he leaned his head sideways to think about it. *"What?"* Dianne asked anxiously.

"Carter knew about the project all along, even before Liza died."

"You think so? Well…he's supposed to be a friend – Isn't he?"

"That is how we have always known him and he loves to stay very close to us."

6

On July 17, 1997, Aduna Gyasi was working at the West lab when the sixth sense buzzed, *"Tutu Babe is on the line!"*

"Hello?"

"Hello Aduna. Man I really need to talk to you!"

"About what?"

"I don't want to talk on the phone though! Could you meet me at the East Lab? It's evening now any way…and you're just half-a-mile up the road!"

"Well," the science wizard offered graciously, "As it sounds so urgent, I think I might close off for the day and head over there."

"Good!" the big entrepreneur replied abruptly and hung up.

In those days, the labs were in Needham Ma., but the cooperation was pushing to have them moved to the heart of Boston. That should have been an impossible dream. When it happened in 2018, it proved that Tutu Babe got whatever he wanted!

Gyasi cleared security at the gate and drove up inside the garage. It shook steadily to the sound of speeding cars on the Massachusetts Turnpike just meters away. He looked down through his window and marveled at how fast the lanes went at that time of the day. The scientist killed the engine and stepped out of the vehicle, heading for the building entrance. He was in no mood to talk to Tutu that day, but he had no option. Gyasi took his sweet time heading up the stairs to the third floor. Before he touched the big metal doorknob, it flew open. Babe stood before him in a white oversized bush jacket that looked like Moses' robe. He waved Aduna inside eagerly. "Come on Dr. Gyasi! I've been waiting!"

The devil's tone told the god-creator that Tutu was about to play an ace! He followed the big man to the main office of the building. At a quarter to six, the entire floor seemed strangely unpopulated. It was missing the usual flurry of workers bustling through the corridors, desperate to close off their long day. Normally they would be busy even at eight PM.

The big man saw Gyasi's befuddled look and offered, irritated, "I gave them all off today and tomorrow! Too many fuck ups last week! I think everyone was tired!"

Tutu Babe seemed impatient. He was panting and laboring forward. They reached the door and the big man paused, resting his weight on its hinges to take a breather. At well over six feet tall, he towered above the little man. Tutu Babe was so fat that he seemed like a walrus. Gyasi offered to help him once, but Babe did not like the part that said *'your appetite for certain foods will automatically change as well'*. He wanted to eat all that he was eating now because they were favorite parts of his life experiences! He loved to eat 'passion' – salt and sugar, blood, flesh and fear! Babe did not care for dieting! Carter James had a plan that would serve him better!

He noticed the question sign still conspicuous on Gyasi's face and responded to it with a sober explanation, "We have problems and I believe you know the answers!"

"I see," Gyasi replied, sounding as confused as he looked. Tutu Babe pushed the door open and penguin wobbled to his desk in a hurry. Aduna followed him into the spacious luxury office that smelled like rich Jamaica coffee. The color purple was loud inside, in spectrums of its different shades, that Gyasi created by brilliant nana technology. Babe's office was more like a lounge. The first thing that caught one's eyes would always be the sweet looking black girl in that dim-lit rainbow corner at the bar and the mauve shaded 'chill zone' facing the windows overlooking the highway. The office desk was near the windows too. Gyasi had to admit, he liked the arrangement, although he would not take as much time to laze around as the big man did. He was too passionate about his work. Besides, he would not want it if it meant trying to own the souls of men, or purchasing abducted girls from third-world countries as slaves.

Gyasi's eyes trailed back to the corner at the bar. Babe watched suspiciously. The young woman smiled at the little man but it seemed artificial. Not the natural warmth you would expect from a Caribbean bartender. "She's new!" Tutu offered defensively.

"I see," Gyasi responded, obviously not impressed. The big man left it at that and waved to the chair, "Please have a seat Dr. Gyasi!"

Gyasi sat down in front of him, "What is it Mr. Babe?"

Tutu ignored the scientist and started up the laptop on his desk. Aduna waited. Babe put a disc into the slot. Then he clicked on a video. The video started. He paused and turned it around for the little man to see. It was Lisa! Gyasi breathed in deeply, but kept a straight face.

"Do you know this woman Dr. Gyasi?"

"Sure I do! She is a good friend of ours who went missing."

"OK! And 'ours' means you and Dianne?"

Babe said 'Dianne' with a familiar tone and the god-creator took note of it. The devil still had useless obsessions that should have passed after high school. No one human being can ever have *everything*!

"Why am I *here* answering these questions from you?"

Tutu gave Gyasi a one-eyed stare and said, "You're here because I can help you, if you will help me. Besides, you're cheating! You've been up to a lot of stuff that you haven't been telling me. You work for Tutu Babe Cooperation, not for Aduna Gyasi!"

Tutu's bluff did not impress Gyasi. The god-creator smiled assertively. "I do my job for the cooperation and I do it well. What I do outside is my business. I had a life before I started doing contracts for you, and I still do regardless. I am not your employee. Neither am I a slave of any sort."

Tutu hissed threateningly, "I don't care what you say! I must know all that goes on with my workers! This is a top security lab! You do not go off on side orders without telling me! I won't allow you to destroy my reputation!" He restarted the video.

Lisa was making entries in a personal video diary. She talked about being involved in a project to put cyber genes into humans. These synthetic catalysts would mutate and improve Man's biological structure. Ultimately, they would become the natural human genetic base. She talked about carrying a cyber twin inside of her, claiming that they were the future of

Mankind. Humankind, she said, was about to blow the cocoon. Lisa talked about her partners whom she called 'Sis' and the 'Sage'.

Aduna Gyasi smiled, wondering where and how Tutu got Lisa's video. He was not worried. Lisa was no fool! He leaned his head sideways in his customary meditative way.

Tutu Babe challenged Aduna with his eyes. Gyasi would not concede! The big man scanned his brain trying to figure on a new strategy. Finally, he retorted all-knowingly, "Mr. Gyasi, you said that she's your friend who went missing. I've got news – which you may already know. She's your friend who went dead! And she did go dead around your neighborhood. Someone is going to prison for murder!"

"I do not own neighborhoods. Tell the police what you know!"

Tutu Babe chuckled, "You know what I think Dr. Gyasi?"

"No I do not. Enlighten me!"

"I think you're the person she refers to as 'the sage'! You have some weird science going on, and you used her as guinea pig. Now she's dead!"

The scientist looked to the bar. "I can think things about you too! I am a scientist, not a murderer – or even a kidnapper for that matter!"

Tutu Babe studied Gyasi's face. He looked to the bar too. Then he shook his head in mocked incredulity, changing his tone, "Believe what you want Dr. Gyasi!"

"That is what I do!"

Tutu Babe was infuriated, but driven to a truce. "Fine! We're all scientists here! Sometimes things happen! Let's change the topic! Let's talk about scientific things you do behind my back!"

Gyasi rose to his feet. "Fine! Things that do not involve some weird attempt to demand! Otherwise I am out of here!"

The big entrepreneur scratched his head. "Alright!" He pointed to the seat and said, "Please sit down Aduna!"

Gyasi sat and waited. Tutu pushed the laptop aside. The god-creator nodded his approval. Tutu Babe smiled cozily. He eyed Gyasi and admitted, "Alright, that didn't fucking go well!"

"No it did not!"

"Lisa was a smart kid."

"I know," Gyasi replied nonchalantly.

Tutu was thinking and Gyasi waited. The billionaire affirmed flatly, "I still have propositions to make. We have no proof of what weird science you do – I'll admit that. Still, if the rumors are true, the outcome could mean billions in profits for the company! So, off the records, talk to me! If you have some incredible stuff going on, I must be a part of it! You need to be careful about what you do with your talent!"

"What do you mean Tutu?"

"Well, for one, I know you're planning to seed Mankind with this cyber-Adam. If it were ever possible that would be a no-no!"

"If anyone planned to do that it would seem like a great idea. Too many bad people hold power. It would level the playing field."

"By giving power to everyone? You think that would be good for world peace?"

"Paradoxically, yes. Not the power in itself, but peace is the sum of equality and justice."

" Bullshit!" The fat man leaned back in his chair watching the ceiling, rocking back and forth. "Your Cyber-Genetics, or whatever Lisa called it, could be a new phenomenon. Can you imagine how it could transform the cooperation and change the world? Making cyber-genes? *Jesus*! That's like a too-good-to-be-true idea! It could eventually lead to immortality, couldn't it?"

"That could only be a relative possibility."

Tutu chuckled, "You mean, 'if anyone planned to do it' – Right?"

"Right Mr. Babe! That is exactly what I mean!"

Tutu's eyes were staring into space. Another question came to his mind and he asked, "You think this sage person could create man or beast from scratch? I mean...maybe that would be impossible...I..."

"I think this person could make flesh, bone and every part of a human from dirt, if he chose to do it. And that created being could live, breathe and reason as other humans!"

"I can't get it! Even if you create a body, how do you create for it a soul?"

"To understand 'soul' we must first know how the unseen energies work. Positive energies define themselves in negative ones. That is the first intercourse, creating conjecture. It is generic 'I Am' consciousness. It becomes human when we program it to interpret particular truths as emotions. This is part art and part science, but the secret of how it works must remain unspoken. The more intelligence the subject possesses the more critical it becomes. Critical intercourse, in itself and by itself, must acquire the 'I Am' factor because reason comes from an object base. A soul is inherent."

Tutu leaned forward, watching the god-creator. His eyes took on the light of enchantment, as if he was daydreaming. He sounded bedazzled when he spoke, "Seeding this cyber might into the masses would be too much power for inferior humans Aduna!"

"Crawling is not too much power for a caterpillar whose destiny is to fly!"

Tutu reflected on Gyasi's words and said, "I have a better idea. Instead of giving unreasonable power to inferior masses, let's use your wisdom to strengthen the power of the state! Our government has contracted Tutu Babe Coop to create the ultimate robot. It will be our own Arpago, the Iron Soldier, almighty and all knowing! Scrap the Cyber-Baby project and get involved for our country! Put your wisdom into creating intelligence for Arpago instead of fantasizing about creating humans!"

"Why would you give omnipotent power to a machine with advanced mental faculties that knows it is not human? Give it instead to a living man! In fact, give it to all humans! The ancient Arpago was a living all-powerful human being from another plane, not an iron soldier. What you propose is a recipe for chaos! I could never agree to it!"

"We believe, by the facts of what we've now discovered that Arpago was an iron soldier! Your idea is more conducive to chaos than mine Mr. Gyasi!"

"My idea will invoke, from the heart of the people, peace with humanity."

Tutu was about to respond to Gyasi's last statement, but he thought about it and decided not to prolong the issue. "OK! You will not do the Iron Soldier! Fine! I'll only ask you to contribute one piece of technology to it."

"And what would that be?"

"You should remember that you work for us, and with that commitment is a responsibility to contribute to our projects."

"That is irrelevant Tutu. I only make what I agree to make and then you pay me. Instead of answering the question you just went off on a tangent."

"Alright!" Babe conceded, "Let me tell you what it is! Create a compact sound frequency weapon for the Iron Soldier! The most awesome acoustic design there ever was!"

"Much of what I created for good you already put to evil use!"

"That's bygones! I need this for the U.S. Army – *Today*!"

"I have no idea what exactly you mean by this request," the god-man lied.

"You know…like making a device that uses sound-waves to kill 'enemies' on a mass scale without damaging *things*."

Babe focused on his desk. He did not hear the wizard answer his question, so he looked up to see him. Gyasi was shaking his head distastefully. Then the little man asked sarcastically, "It will kill 'enemies' and not 'people'?"

Tutu began to answer before realizing the rhetoric in Gyasi's question, "Yes…it…" then he paused, refocused and surmised. "I take it that you're not interested in this one either. It is to protect our country you know?"

"We have the capacity to protect our country without annihilating, or interfering with, other races and tongues! Propose a better project to me and I might be interested!"

Tutu Babe's jaws tightened as he responded in an aggravated sneer. "My next proposal is this! Create cyber-genes that won't reproduce or transfer from one person to the other! Otherwise, the product will be lost to us!"

"Elitism is not my style Mr. Babe."

"Are you refusing me on all counts?"

"You went against my rules on all counts Mr. Babe."

"Too bad for you! Kenneth Groh is the most able replacement! He begins the Iron Soldier Project on August 31. I am the only man with the authority Gyasi. Without me, your agenda is dead! Consider your cyber-baby project dead as well. You will not be able to put one foot in front of the other! And when you are ready to grovel, you will come back here!"

Gyasi rose and headed for the door. "Aduna," the fat man called soberly. Gyasi froze and half turned to face him.

"Your cyber-baby project would include the twins in Dianne's womb!"

Gyasi's heart stopped and he felt like fainting. "*What?*" he croaked, trying his best to hide his utter shock and frustration! The devil knew it all! Gyasi saw it written all over his face! This was it! It would be hell to save the unborn!

7

August 27, 1997. At 11:45 PM, they delivered the twin!

Aduna Gyasi was in the West Lab with Carter, finishing his personal robot. The intellectual faculties of this model surpassed that of the Iron Soldier, but Carter would not know. To him, it was an ordinary housekeeper, of which he needed not tell Babe. It was the size of an average person and seemed harmless as could be. This robot would reason conscientiously as any human, but only ten times more logical. Aduna created the machine to guide and protect his unborn child, as keeper of the cyber-baby project. The machine could destroy a small army, but it could not conceive the fortitude to go against the order. The wizard programmed it to act only on the command of Aduna Gyasi and Anne Eve Gyasi, who was still unborn.

Had he been a step ahead, he would have created this entity to keep his dead children alive! It never happened, and he had serious issues of inadequacy and guilt to live with.

Gyasi smiled at Carter, who was always with him, thinking he did but not knowing much! Sometimes, even with friends, one had to keep it that way! Ignorance could protect the subject from betrayal and it could protect a friend from dishonoring himself.

They finalized the robot and sprayed it with instant dry. It looked like sterling silver. James Carter brought it upright and they observed the thing with pride. "I know what you should call it mate!" the tall Englishman said with much assurance.

"What?" Gyasi asked indulgently.

"Doo-Bug, the élite of all domestic robots!" Carter chuckled and then added, "Which Tutu Babe will never know we did in his lab!"

Gyasi grinned cynically. "I know partner! The name is Doo-Bug!"

"Great!"

Gyasi's cell phone rang, but Carter could not hear it. Nurse Hudson was on the line! Gyasi jumped at the sound of her name. He took the call. Carter watched the little man talking to himself. When the call ended, Gyasi turned to his partner, "Let us grab this robot and go Carter! This is it! With Dianne I mean!"

Carter froze in anxiety. He was not expecting it so soon!

"Hey pal, you are freezing up on me! Let us go!" Gyasi repeated urgently, bringing his friend back to awareness.

"H-How comes? It should be t-two days from n-now!"

"I was wrong! It is two days early!"

"Bloody hell! I cannot conceive you being...*off*..."

"You have *seen* it! Come on now!"

In ten minutes, they drove through the pine trees to the house at the center of the property. Carter jammed his forehead against the passenger window to admire Gyasi's little forest. The white mansion sat at the highest point on the land, which gave it a majestic appeal. Carter loved it now! In the beginning, he was against Gyasi building so far out in isolation.

Gyasi pulled up to the driveway and parked the old VW Jetta. He hopped out of the car and fumbled for his remote, while swearing sarcastically underneath his breath, "Hell I had to go to that godforsaken lab today!"

Carter got out and pulled the back door for Doo-Bug. Then he caught up with Aduna and joked, "At least Doo-Bug and your twin will have the same birthday! Less hassle for you!"

Gyasi remembered the robot and turned to see behind him. It was hurrying to catch up. The god-creator smiled at it patiently. "Take time to learn my friend! You will be less confused as time goes by."

The door came open. Carter hustled into the living room ahead of Gyasi. Doo-Bug stood a couple of feet behind them. "Should I come in?" the robot asked politely. Gyasi did not answer, but Carter James hollered from inside, "Yes *fool*! Hurry! You are a bloody computer! Stop asking stupid questions!"

"More often than not," the robot asserted wisely, "the answer is stupid and not the question. I'm taking my time to learn. I'll be less confused as time goes by! You on the other hand, haven't learned much in many years of time – Have you?"

Nurse Hudson was in the room with Dianne, who was moaning in pain. Aduna hated what he did but it was necessary! Carter pushed the door hurriedly and went in. Dianne was fighting to stay calm. A human child lay beside her. It was not breathing! Dianne was sure that death would come to her at any minute. Gyasi stepped up to her side with Doo-Bug behind him. Dianne tore her eyes wide in disbelief and half-whispered, "What did you do to us Aduna?"

Nurse Hudson had something in her hands. It almost looked human but it was only half-visible, like some kind of jellyfish. It struggled too, but not to breathe. It was having problems adjusting in this dimension!

The nurse moved to Gyasi's prized resuscitation chamber, methodical as the true professional she was. Not even the wide-eyed confused robot was as calm as she was. Carter took up the sibling and hurried behind the woman. Gyasi charged off for the chamber too, crossing paths and colliding into Carter's back. The tall scientist did well not to drop the child, already battling for life. Gyasi's eyes trailed across to the Nurse. She had seen!

The medium built blond woman was too strict on principles to care about who was boss. Sometimes it made her seem snobbish, but she was usually right!

Nurse Hudson closed in the strange being, grabbed the human child from Carter and closed her in as well. Then she turned to face the god-creator, stern as a rock. "Please step out of the room Dr. Gyasi!"

Even in her pain, Dianne nodded in agreement. Gyasi turned to go, hesitating to see if Carter was not coming. Then the nurse said, "Dr. Carter! Please help me with Dianne!"

The little scientist moved to the door dejectedly. Doo-Bug watched with emphasized curiosity. Nurse Hudson noticed the robot and waved it off too. It wheeled around to follow its maker. "I shouldn't be in here either! And I'm not the clumsy one!"

Gyasi wanted to stay and see, but he restrained himself. Whatever happened he would not resist the wisdom of his faithful. His heart sank inside of him. Dianne faced serious complications, but he had planned for them. Her life was in jeopardy, but Nurse Hudson knew what to do. Failure was unacceptable this time! She had the best equipment in the world on hand. He could do nothing that this meticulous nurse would not do. He already wrote the script. Now it was time for the sequel to unfold!

He went to sit on the couch in the living room. Something was underneath his thigh. Gyasi eased up to see what it was. The thing rolled between the cushions and he reached down to pull it out. It was Dianne's hand-held cell phone. He took it up and cleared the screen. There was a text message from Tutu Babe! The little man's heart leaped to his throat! Why would Babe send text messages to her? They knew each other a long time before but only as schoolmates. He was the only one in the house who ever talked about Tutu Babe. This was scary indeed, considering the big man's threat!

He could not help it! Tutu Babe was his business associate and archenemy all in one. He opened the message and read, *"Congratulations Mrs. Gyasi! I hope you enjoyed the chocolates!"*

Gyasi jumped to his feet, feeling his heart tear mercilessly against his ribcage! At the same moment, Nurse Hudson came out of the room with Carter behind her. She went to him. Before she could speak he asked, "Did Babe send chocolate for Dianne?"

She stopped in front of him stunned, dragging her busy brain from the problems at hand to the seemingly trivial. "Sure! Twice per week for the past weeks!"

"Why did you not tell me?"

"Your wife would not. Why would I? Besides, she didn't need a scientist to help her handle herself! It's not like building a feisty robot!"

Nurse Hudson's logic as usual! He paused to think and then began, "Has she been..?"

"Eating them? She loves chocolates, but if she was a fool she'd be married to Tutu Babe, Dr.Gyasi."

Gyasi sighed, mustering the courage to ask, "How are we doing?"

"We did all we could!"

It was near midnight now. Gyasi walked by Nurse Hudson who watched him anxiously. He went into the room and closed the door behind him. Nurse Hudson turned her focus to Carter. "I guess you should go ahead now!"

Carter James took up the phone and dialed Tutu Babe. Before it rang the third time, he heard eager shuffling. "Hello Carter? What's the good news today?"

"We just delivered the twin!"

There was a pause on the other end of the line. The boss did not like what he was hearing. Finally, he hissed, "You told me it would not be for the next two days! We fucking planned the whole thing to work th…"

"Gyasi miscalculated Mr. Babe."

"Bullshit! Aduna Gyasi is never wrong!"

"He was just as surprised as I was!"

"I see. So what now?"

"There's no worry! He failed! One is dead! The other will be dead soon!"

Tutu Babe exhaled. "*Great*! Whatever you did on short notice was a good start! But what does 'will be dead' mean?"

'*Whatever you did*'? The scientist chose not to correct Babe. He did nothing but if he took the blame for their lives that would translate as more money in his pocket. "I mean that it is in a very critical condition. It cannot possibly live for more than a week."

"How sure can you be?"

"A hundred percent."

"You're speculating! How can you be a hundred percent sure?"

"I'm bloody sure! Apart from Gyasi, I'm the best biologist in the world!"

"Aduna Gyasi was wrong about their birth date today, and he's better than you!"

Carter fell silent, thinking and hoping the big man did not lose patience. He managed to speak before Tutu began to pry, "Mr. Babe," Carter affirmed in a husky tone, "If I'm wrong I'll bloody *fix* it then!"

"That's what I wanted to hear! You're the man! Never let me down!"

"I know. I'm the man!" Carter returned weakly, feeling like the slave that treads the winepress only for the benefit of others. Tutu Babe sensed Carters thoughts and chuckled gleefully. Then the big man changed the topic. "How's Uche doing? Is he learning well from my friend Aduna?"

"He's impressive Mr. Babe."

"You're the man!"

Carter James hung up the phone. Nurse Hudson applauded in her ignorance, "Brilliant Dr. James! Absolutely brilliant!"

The lanky man looked unsure of himself, but managed to gather his wits to take the credit, "I know! We worked so hard to bring them into this world, only for them to be dead!"

"It sucks – Right? Losing a wife and kids all at once! Poor Dr. Gyasi!"

Inside the room, the wizard smiled spitefully. The enemy had served him lemons, so he was going to make lemonade!

Tutu Babe sat in his office twiddling his thumbs. The phone rang. He saw Gyasi's number and picked up. "Hello buddy!"

"Have you been sending these chocolates to my house?" Aduna questioned hastily.

Tutu chuckled, "Dr. Gyasi himself! Sure, I send chocolates for her. Why?"

"You know why! You killed them!"

"Are you insane? If someone took the bastards out, it's good riddance! You want super-human babies? Talk to Tutu Babe! And my chocolates are for Dianne! They're perfectly safe!"

"You killed Dianne too!"

The big man almost toppled over backwards on his chair. "What the hell did you say Gyasi? Dianne…dead? That's not possible!"

Babe sounded confused. He had a hint of panic in his voice but Aduna knew what it meant. People like Tutu Babe lived too high above the law to worry about legal consequences. That devil feared for Dianne's life…or he feared her dying before he could prove himself the winner. He was that disgusting! He always got what he wanted. She was the blemish on his record. That needed to change! For her to die would not work for him. "I don't believe she's dead at all Dr. Gyasi!"

"You want to let your Police boys come over here to check?" Aduna agitated.

Tutu Babe paused. He did not know Carter's methods. He only paid for results. If it meant killing Dianne too, he would pay the man and live with it! What a difference a few minutes made! When Carter called, only one child had died so far!

"I don't know how that shit happened Aduna!"

"You threatened me!" the scientist snapped feigning suspicion.

"I did not threaten you, and I certainly did not threaten Dianne! I threatened the little bastards but I could not get to them! Thank god one died! Hope the other dies soon!"

Gyasi waited. Tutu was bound to take the initiative. Ironically, Carter described Tutu as one who would kill your mother and expect you to say 'thank you'. Finally, the big man asked, "Are you there?"

The wizard did not answer.

"Gyasi?"

"Yes…I am here!"

Tutu Babe was on the offensive now. He spoke, calm and decent, as the devil he was. "Don't judge or accuse me Gyasi! You murdered Lisa Stare! I'm innocent!"

"I was not judging you Tutu!" Gyasi said matter-of-factually, "You made a threat! I had to call you!"

Tutu was breathing hard. He spoke again, sounding like the real Tutu Babe. "OK Gyasi. Fair enough! Here's what will happen! I'll forget you even made this call and we'll get back to normal. You have a lot of work to do in the lab. I'll even grant you the favor of still keeping your secret about Lisa!"

Aduna Gyasi hung up the phone while Tutu felt like the winner. He walked to the side of the bed to see Dianne's still form. It reminded him of that fateful day when Lisa Stare breathed her last. He had to convince Lisa. Dianne chose it for herself!

He went to the resuscitation system to see the languid newborn. "Oh Dianne!" the god-man lamented, "I told you to end the project! It is making me suffer so! Now you have to be dead and they robbed me of my only child! Tutu Babe is a tormented vampire! He requires blood at every turn. Soon it will be bitter for his belly!"

8

May 10, 2020.

It was the long-awaited day for the Iron Soldier Press Conference and Unveiling. Tutu Babe chose the state of the art East Lab Assembly Room 02 to stage the historic event. He built the venue two years before, near Harvard Square and it seemed like the ideal place for the Boston gathering.

In-house technicians busied themselves doing final system checks across the huge conference area. The media turned up as early as 7 AM, although the coop scheduled the event for 9:00 o'clock. Half an hour before the start, they swooped down on the coffee stand sponsored by Boston Way Burgers. Eager workers kept on their busy way, balancing last-minute chores with life-saving coffee and muffins.

The big day had come, intended for jubilation among the adversaries. Tutu Babe would finally crush Aduna Gyasi's head! Instead of conforming, the science wizard had to prove himself a prophet among god creators! They would expose him as the imbecile! Every scientist who had any wisdom at all assisted the coop to create the perfect model.

No one dared contradict Tutu Babes authority. He funded both labs – East and West. Every scientist who excelled did so under the name of his cooperation! The press loved the big man! That was why it hated Aduna Gyasi! Whatever the press hated, the people loathed with zealous fortitude.

By 9:00 AM, all journalists and invited guests were present. Every paused on air to exhale, ready to experience history in the making! Babe declared this conception as 'the closest humans would ever come to creating Arpago', the ancient god-man from the heavens. He claimed that it was the absolute cyber-being. The only one who begged to differ on that affirmation was the ever-contradicting Aduna Gyasi as usual!

In 2020, it was ludicrous to argue that humans had greater faculties than machines. Not with the discoveries that even Gyasi made! Tutu interpreted the wizard's work as proof that humans could not derive almighty faculties! Gyasi claimed that it proved the opposite! He was the scientist but Tutu controlled the media. The press reported that the created legion was all-powerful. Master-humans would own and control it but they could not equal it! Gyasi's contradictions turned to condemnation.

When the point man stepped to the podium with the remote in hand, jaws slackened and mouths fell ajar in anticipation. "Ladies and gentlemen – Members of the press and other invited guest, good morning to you all!" Groh greeted them with a huge grin and a smart bow. "As you can see, it's about 9:00 o'clock now and Babe Coop did not send me here to give a speech! They gave me the privilege to open a mammoth chapter of history to you! This is a giant leap for humanity, and a mega fail for our detractors! Without further ado, let me do what they sent me here to do! With the press of one button, the most powerful entity on earth will come to life, but only to the service of humankind!"

With hearts beating fast the audience focused on the nine-foot tall iron monster at the podium, to the left of the scientist. Its breastplate was red and blue with little white stars shining throughout its length and breadth. The rest of its body was metallic gray. Arpago was an imposing herculean looking beast. However, apart from its stature, there was no way to tell that it was lethal by looking at it. Neither were there any signs to show that it bore arms. The media staff was anxious for it to awake.

They wanted to see the demonstrations of what made this thing more potent than any army on earth was.

Dr. Thomas Groh hit the remote. The Legion's eyes popped open and the room fell silent. "Do not get too excited yet ladies and gentlemen! It will take a while to booth! After it is up, we will begin to do the demonstrations. You will get to appreciate the might of Arpago! I'm telling you, it goes way beyond your wildest dreams!"

He stood to the side of the nine-foot tall machine and pointed. "Notice the eyes! They are glowing with that dull multicolored light. When the Iron Soldier is ready to focus, the eye color will change to 'sky-blue'. However, underlying that color is what we call 'target red', which you would see if it were in attack mode! With this great creation, the United States of America has finally achieved the highest degree of world domination since the beginning of time!"

They watched the monster do a range of motion, starting from its neck down to its legs. Then it stood up tall, its eyes turned to target red and it declared with inhuman ease, *"Humanity is the highest threat to the United States environment, Dr. Groh! The Legion will create and put a solution in effect!"*

They heard it talk and awed over its faculty. Most of them applauded, happy to see the demonstrations begin. It took a while for the implications of its words to get through to the scientist! No one expected an 'error' so soon! They remained delightfully enchanted for that delaying instant. Then Groh's brain picked up and he exclaimed – speaking for everyone else, "What duh *fuck*?"

The scientist's outburst was the unquestionable prescription to panic, but the unexpected twist to the proceedings took them by surprise. They were hard-pressed to react.

You could stall men for answers, but not robots. The Iron Soldier had no obligation to acknowledge the insignificant being's question! In its iron cranium, Groh was a mere enemy of the United States! Machines begged no apology and entertained no debate. That one pronouncement from its mechanical lips was final! It echoed around the room for all and sundries to hear. They gathered their thoughts and translated it. *Extreme* fear!

Again, Groh uttered on behalf of the souls in the assembly hall, "*Fuck* me!"

Groh's symbolic act of prostration raised the hair on everyone's head! His undignified words went for all too! Thomas Groh would know the necessities more than anyone else could! He realized his mistake and then struggled for self-control. Groh was *'the Legion's'* superior! He should command it, not bend over for it! "Stand down soldier boy! Stand down *now*! That's an order!"

The Iron Soldier refused the command! It turned to face the assembly and Groh could see the apocalypse! Humankind had no moral authority to deny the Legion's logic. It was futile to consider any weapon of defense. This was *it*!

In one heart-stopping moment, all conversation transformed to telepathy. Terror held strain with a suffocating effect! The guests gagged on invisible fear, with all eyes bulging from hollowing sockets. They seemed to move in slow motion. The East Lab Assembly Hall would become a death chamber! After what seemed like an eternity, someone managed a premature scream. A three hundred pound man in black headed for the door, paddling like a moon-walker and forcing his fear-frozen legs forward. The floodgate of frenzy opened!

Half the people in the area braced themselves, surrendering to what was coming. The other half made desperate attempts to escape, all odds stocked against them.

There was never room for failure, considering the stakes. It was time for a glorious exhibition. Instead, adverse hell was coming! They should have discovered such a grave error in the East Lab design of artificial intelligence long before!

Then again...maybe every fault had been exposed beforehand. No one wanted to listen to negative comments. Now, none of them wanted to face the result either. They made Aduna Gyasi a solitary, even a byword, for trying to point out their mistakes. Groh and the others were so damn sure! Even the media took sides on it!

All eyes, except his own, questioned him for the queue. He was the professed genius, and point man behind the project. Besides, Tutu Babe was nowhere in sight!

As he thought about Gyasi, Groh found a ray of hope for those not yet in Armageddon's way! It was a no-brainer to choose between death and a taste of one's own vomit. He mustered all his strength and shouted "Gyasiiiiiiiii!!!"

Although the outburst sounded like a desperate prayer to some god, it was the best answer to the impossible question. The wizard was not there, for he was not welcome! Nevertheless, Groh knew that if any man on Earth could challenge this thing, it could only be Aduna Gyasi. He was the only cyber master who refused to help the cooperation create it! Even if he were there, defeating the Iron Soldier would seem impossible!

One reporter took a desperate look at Groh and understood. She saw the Legion raise its hand to the scientist's face. There was a flash like lightning. Screams from petrified journalists and company were overwhelming. Groh's head disappeared, and vapor rose from the stub of his neck. He took time to wobble and make a wavering step forward, before slumping to the floor. She fell flat on her stomach too, shivering uncontrollably. Every time the Legion raised its hand, lightning came with a slight crackling sound, then a sizzle, followed by vapors, raising the carnal stench of disintegrated flesh and bone. Mostly, the heads alone disappeared.

Souls were running out fast – from wherever they were hiding, under desks, behind columns, beneath other bodies…or hugging each other for comfort. Her number was sure to play soon! Thank God, the beast was in no hurry. It concentrated mostly on the ones that made it to the door.

Amidst all the screaming and dying, she reached for her phone to dial Aduna's number. He was the man she helped to shame! He would not be too eager to hear her voice!

The Legion's synthetic mind calculated the variables and its thoughts appeared in writing on the overhead screen that Groh set up for demonstration. On the opposite side of the podium, a printer printed the legion's activity report.

Mission: Annihilate human threat.

Act on authorized intelligence=Active

Mission Status:

Current location=Massachusetts Avenue. East Lab assembly/East Lab Complex

Resistance Status=0%

East Lab human count=107

Number extinguished=15

Number of survivors=82

Next target=Harvard University.

Total survivors on campus=24385

Basic Command Status:

Act on command=Scan for survivors to extinguish

Human threat level=Code Red=100%

Threat category=Environmental

Worldwide human count=7,093,569,786+3+5++++

General Command Status:

Extinguish worldwide human population

Estimated time to completion=35,040 hours

Estimated resistance time=10%

The Legion Status=Active

Access to Satellites=100%

Universal control of celestial resources=100%

Extraterrestrial contact=1 Watcher Mother Ship+98 sub crafts

Extraterrestrial Definition=Angeloid Observers

Known Celestial Forces=Angeloid Observers=Passive Status=No Threat

Iron Soldier Mandate=Serve and Protect.

Legion Prototype=Omnipotent Warrior

Iron Soldier, USA.

Authorization=the United States Government

All-knowing-programmed faculty=Active

Manufacturer=Tutu Babe Cooperation for the US Government

All rights protected under United States of America Laws

*Patented and protected=**No other prototype to ever exist***

*Patent Protection Status=**No other prototype attempted in the world***

*Attempted Breeches=**1**=**Russia attempted to access information**=Nullified*

Successful   Breeches=0=Legion   Prototype Protection=Active

USA Supreme Status=All Earthly Sovereign Subdued

Threat Level=0.01%

Celina watched the gigantic robot, seemingly frozen in place, as it allowed its programs to update. Even as it seemed frozen, it acted to pick off and silence anyone who tried to make it through the door. Groh had explained that it would take a while to update, as he had just activated it. It only took seconds to malfunction!

The phone started ringing, and her heart raced even faster, compounded by anxiety in the audacity of hope. She heard the line open. Then, she looked up. The Legion knew! It turned its focus to her, about to act! Celina saw death coming! She would not survive, but she did the last honorable thing for the rest of humanity. She took her last moments to give Aduna Gyasi the hint.

"Legion! Help! Gya..."

....

It was not an official day off, just a day when he would rather not be at work. Monday the 10th of May promised to be very cold in the East Lab! Groh would unveil Tutu Babe's monster! Everyone knew that Gyasi was totally against the idea. Now they were going to prove him wrong. They thought...

He hated idle days like this, when he had to reflect on 'consequences'. His children were gone! Some things he could have done differently! He was stubborn but he weighed personal priorities well, against that of a greater good. Aduna Gyasi was the supreme earth wizard. Elite scientists across the globe referred to him as 'God'. Now…as he was 'God', he had questions to ask of himself. What would it take to bring his beautiful children, whom he had made dead, back? He was living in perpetual pain!

He flipped the television channels until he became bored of it. Then he rose lethargically and walked to his study. The notes he was working on were still on his desk. Gyasi sat down, fumbling for his reading glasses. He soon gave up!

"Doo-Bug?"

He hardly heard a sound as the beanie-eyed robot wheeled itself into the room. "How may I help you Sir?"

"Find my reading glasses, please!"

"OK."

Doo-Bug about turned and wheeled off in the direction in which it had come. The little scientist relaxed in his chair and waited. He pulled his beard thoughtfully, as an idea came to his mind. Maybe he was 'too negative' as they said, but it was only because of the gravity of the situation. If his theory were right, what would happen when they activate that demon? He saw the contradictory way they created the programs. How could he possibly be wrong? Besides, he had never erred before! Well, he had secretly done all he had to do. 'Due diligence' he would call it!

The robot re-entered the room with the reading glasses. Gyasi reached out and took it.

"Thanks Doo-Bug. What would I do without you?"

"Without me sir, you would search around, not find your glasses, and be miserable!"

He was about to say something, but paused to muse at Doo-Bug's impudence. He redirected his thoughts. "Where is Weasel?"

"He's in his room watching the Discovery Channel."

"Why? He has work to do. Tell him to get his butt over here pronto!"

"With pleasure sir!"

Doo-Bug and Weasel had a perpetual battle going on. Weasel was not a robot, but he very well might have been. Between him and Doo-Bug, Gyasi did not see the difference. He needed them both. He could hear Doo-Bug shouting out the message, deliberately and in his own words too!

"Metal-headed asshole! Your boss said to get your lazy butt up and to work, before he fires you!"

Sometimes he was certain that Doo-Bug had *feelings*. Maybe it did too! He had gone a bit far with his secret *'Persona Code'*. That was how scientists wondered off to great abominations. Scientists like Groh who took orders from business moguls like Tutu Babe.

They wanted Aduna to give his human genius to that beast they created. No way! If they were to create a beast, that beast had to represent human evolution.

Weasel charged into the room panting, "Yes Sir – Dr. Gyasi Sir?"

He was a tall, thin, deformed character. Weasel had a straight face, with a long crooked nose. His bottom jaw was out of line – Way out, exposing his bottom teeth. His left arm was half the size of his right one. The same went for his legs, except this time the right one was smaller. He struggled to do the simplest tasks, but Gyasi meant to transform him eventually. Weasel was a major project for the god man.

The little scientist was about to respond to the 'apprentice' when the house phone rang. Gyasi stared at it suspiciously. It continued to ring, and Weasel seemed confused. The 'apprentice' was about to ask, "Sir, are you going to pick that up?" when Gyasi nonchalantly took it off the hook and placed it to his ear. "*Hello*?"

The master's face went pale and he gently brought the phone down. He could feel Weasel's eyes on him.

"Weasel," he croaked.

"Sir?"

"You remember that container I created at the West Lab site?"

"Yes Sir!"

"Bring it...and the gun..."

"You mean the gun that doesn't fire anything?"

"Yes. It is time to test it."

Weasel about-turned and hurried off. Suddenly he stopped at the door. "But Sir, if it doesn't work, what's its purpose?"

Gyasi ignored Weasel's logic, "Put some water in it, and tell Doo-Bug to get inside the bus. Hurry!"

Within a couple of minutes, the three were charging through the front door to the parking area. Gyasi went as fast as he could and Weasel found it difficult to keep up the pace. Every now and again, the scientist flashed an anxious look behind to see how far he was. He hobbled along, doing the best he could, one leg being shorter than the other was. Doo-Bug knew how hard Weasel was trying, but he had to shout, "Come on hop-scotch! Can't you see we have an emergency here?"

The robot hurried to the side door and threw it open for the master. Then he went to the front and took the driver's seat, "Where to, boss?"

"To the East Lab Doo-Bug – Get there since yesterday!"

The Robot held on a second for Weasel to pull himself in beside the scientist. "Glad you finally made it!" Doo-Bug quipped sarcastically.

Gyasi looked around, making sure he had everything. "Where is the Ray Gun?" he asked urgently. Weasel handed it over.

"Here Sir. It looks like a water gun for kids!"

"Yes it does!" the scientist agitated, "The only thing is that it is empty! Did I not tell you to put some water in it Weasel?"

"Oh! I forgot to say...I thought it would take too long, so maybe we'll fill it when we get there!"

"*Shoosh*!" Gyasi vented in frustration. He handed the gadget back to his assistant, "Here! Urinate in it!"

"What sir?"

"Weasel, if you do not piss in that thing now, I swear I am going to pulverize you!"

Doo-Bug poked his head around for a second, "Piss you lame bastard!" he interjected, "Or be pulverized! Too much metal in your food! That's why you're so ignorant!"

9

The smart-looking Commander Bayfield turned to face the line of officers. Then he spun sharply to continue his deliberate pacing back and forth along the line of subjects. "Gentlemen, Digital Navigation Rifles and Hand Guns are very expensive, and their bullets cost ten times more! They designed them specifically for Marines that had to go on rescue missions overseas. Digital Navigation Units proved very useful in helping us keep the edge in the war against terrorism! The FBI and the CIA immediately decided to include them in their armories as well."

Three youthful marines stood off in the background to the side of the commander. Two of them carried regular handguns. The third had a handgun too but a peculiar rifle, resting upright against his left leg, drew the eyes of the police officers.

Bayfield was a strong built soldier; the most rugged looking customer one would ever see underneath a beret. In just a couple minutes, the hot sun transformed his pale face to red and a bead of sweat rolled down into his left eye. He brushed it away with his right knuckle. Then he turned casually to take a spiteful peek at the blazing sun in the open blue sky behind his barrel frame. The police officers watched with incredulous zeal. Like he could do something about the weather!

Bayfield turned to continue where he left off. "Now, finally, after two years of bickering, they are going local; meaning they have decided to afford state police a limited amount of these very expensive weapons! I get the job to introduce it to you! I need not stress on the need to take good care of these gems. That's not my department! I'm with the Marines! I know that there will be consequences for any irresponsible actions surrounding the use, abuse or neglect of DNU's; rifles, guns, parts or bullets!"

"Let me emphasize that the Marines do not have time to waste on exhibitions for police officers! I'm still wondering why I'm here! I'm going to give you fifteen more minutes of Marine time! Make sure you understand the demonstration fully in that eternity! Keep your eyes on Hadley, Shane and Jarrett on my left!"

The Commander walked off without another word. Alvin chuckled and Sam shook his head in disgust. These military boys were arrogant wise asses! "OK, so we'll just watch the three stooges then!" Halloway quipped from behind them, and the thirty odd officers fought to contain their laughter. The soldiers went about their business oblivious to the resentment.

Hadley stepped in front of them holding a DNR aloft. He turned the rifle so they could see on top of the barrel. It had a digital device, much like a camera, which showed the image of whatever stood before the nozzle. Shane and Jarrett walked off to about twenty meters in front of the police officers. Then Jarrett stepped behind his colleague and put a gun to his head to imitate a hostage situation. Hadley turned to the police officers. "As you can see, when I point this weapon in the general direction of the hostage and villain, I capture the image on top of the barrel here! The trick about this weapon is not about your aim! As long as you can see the image on your digital camera, all you need to do is mark your 'X'! We can hardly see the culprit behind the hostage but our rifle will fix that problem! We're all snipers now!"

That thought impressed Halloway and he whispered sincerely, "Let's see this one!"

The soldier saw their unbelieving faces and smiled, "OK! Some of you won't be able to see this from where you're standing. That's fine! You will get to handle this weapon and do it yourself later on!"

The Marine touched the menu button on the digital screen. It made a beeping sound, and he continued in a satisfied tone, "There we have it! Aduna Gyasi's 'X-ray Photo Match Technology', courtesy of Tutu Babe Cooperation!" The screen was asking, '*What number layer do you require?*'

The soldier explained, "This technology specifically detects and photo captures human forms from before or behind other objects or humans. It also identifies and differentiates between these objects. Then it automatically downloads the image to the grid to gain info on the suspect, including his last phone call. Everyone who ever walks on the street has grid pictures. It matters not if he covers his face all day! When we load a hit to the DNU, intelligence automatically downloads to the miniature missile. This is awesome stuff! It is the same technology used for the anti-theft motor-vehicle systems today. You'll remember that it wasn't working well to scan wallets – I mean, what if the ID in it did not belong to that person? Well they invented the photo match system to solve that problem. That was before Babe Coop came up with the DNU idea. It also solved the problem of identifying hooded robbers, as you will see in a moment."

"If up to ten humans were standing in a line – one behind the other, the rifle cam could capture the tenth person at the back. Just as it will capture our villain Jarrett in a second! It is potent from a range of up to five hundred meters!"

He signaled the cops to view the camera again. Jarrett was still standing behind Shane. The soldier grinned and said, "You are looking at layer one! That's all we can see normally. Let's hit the button for layer two and see what happens!" He hit the button and there was the shocker!

Alvin looked at the camera and then at the 'hostage' in front of Jarrett. Shane, 'the hostage', was still there but he was missing on the photo capture! All he could see was Jarrett pointing his gun at thin air!

"Bombo-claat!" Sam muttered in a gravelly Jamaican accent.

"Good!" the soldier quipped satisfied. "We were all paying attention! We have our villain here, and our victim is invisible. That's how we like it! All we have to do now is press the target area on the villain. Let's go for the head. Touch the screen, in this case, between the eyes and…" The soldier did the touch and the machine-made a ticking sound. "Now it's locked on! No need to worry about your aim! *This* is the secret!" He reached into his side pocket and came up with a Digital Navigation Bullet.

"This is a guided missile. It automatically locks on as you touch the target on the digital screen. When you pull the trigger, as long as you point in that general direction, this super bullet is going to explode inside of the target. In this case, it will shatter the enemy's brain; and it won't harm the hostage either! Don't worry about how it's going to get there. Straight, zigzag, cutting corners – whatever! These little gill-like sleeves – I hope you can see them – are like little wings to navigate and cut curves! Navigation Bullets are ten times heavier than ordinary ones. If you are the selected DNU shooter, you will not receive more than a few!"

Hadley replaced the bullet and took out a handgun. "This is the hand gun of its kind – a DNHG. It works by the same principles. Before we look at it, I am going to show you the wonders of Navigation Units! I should also tell yo…"

The Commander came back with the police chief hurrying behind him. "Boys, you have to break! Hell is popping in Harvard!"

…………………..

The Iron Soldier reviewed its tally. One hundred and seven people were dead in the East Lab! The legion was now at fifteen percent of full capacity. That was enough to annihilate the subjects in the area and destroy the National Guards when they came…for they would come! The machine intercepted and deciphered every telephone call and radio message in the state of Massachusetts.

It was ready to go to the street and destroy three dozen Police Men waiting their turns. Aduna Gyasi arrived just in time to see the mighty creation burst through the East Lab wall as one would rend a piece of paper. It took to the pavement attracting a hail of gunfire, mostly from high-powered rifles. The bullets could not pierce its invisible energy armor. Their resistance was futile, but there was nothing practical to try. Luckily, the Legion was still on low power. It picked out its targets one and two at a time. Nothing could save them; not even the armored vehicles behind which they were seeking cover. Each time it fired, someone's head disintegrated and vapor rose from the remaining stub. The air smelled like burning flesh!

Halloway saw the old man jump out of the bus with what looked like a toy gun! The cop could not risk getting across to him. "*Fucking civilians!*" He cussed underneath his breath. He indicated to Sam, who was in a more advantageous position. Sam frowned at him, *"Wha'?"*

Halloway pointed to Gyasi. Sam nodded, and Alvin turned around. "What does Halloway want?"

"Look over there Al! One ol' man coming with a raas-claat toy gun! No trouble him! Make robot explode him bombo-claat ol' head!"

Alvin looked and saw Gyasi. He feigned a laugh, and then spoke between clenched teeth, "We're all gonna die with him anyway!"

Sam caught his eyes, not knowing what to say. Alvin was just telling it like it was…but he resolved to try until he died!

The pelting sun glistened from the shiny form of the Iron Soldier, which seemed to add to the terror. It gave the illusion that its body was generating the heat of the cosmic that dazzled from the pavement. Blood was gushing now like Armageddon's flood, and the smell of it was nauseating. There was no question about global warming in those days. The people accepted the fact that times changed according to cycles. It was the way of perpetual motion. The high temperature seared the blood on Harvard Street. It thinned and ran down into drains and crevices, only to thicken wherever it stopped.

Halloway felt like a sacrificial lamb in the battle. How foolish the hopefuls were – including him! They saw the monster of death, and they knew they could not defeat it. What was the solution? Go out and sacrifice their sizzling heads to it, one man after the other! They were all in line, like products on a conveyor belt! He gritted his teeth, desperately trying to numb his mind to the experience. *"Jesus Chris'!"*

"Yes!" Sam reiterated with emphasized disgust, watching the brilliant body before him. "Jesus Christ!"

Gyasi stood almost beside Sam now, pointing the thing at the Legion. The Iron Soldier analyzed all the threats. It was taking out the most dangerous ones first; although they were all like zero threat to him. There was a man with a gun filled with liquid containing ammonia. This one was the lowest risk. The Iron Soldier fired and took out a cop six yards behind Gyasi.

Gyasi steadied himself and got ready to shoot. Then Sam charged, pushing him to the ground in a flash, "Get down old man! Get out of here! And put away that piece of shit you lead head!"

He had knocked the wind out of Gyasi's lungs, and the old man laid still, even after Sam rose to pull him into cover and get back to business.

Now Gyasi was in a sandwich between the two cops. He tried to go back out again, but Alvin held on to him. "I see you're going to be a problem! Sit down!"

Alvin dragged Gyasi's shirt and the old man fell to the ground. Now he was out of the circle! What could he do?

Then Halloway took a dive to change position and fired off a couple of rounds. The Legion turned to spot him out. He was hiding behind a patrol unit, which they knew would not help! This thing was running a hundred percent on target! It locked on to Halloway! The big cop ran to a line of seven cars, his heart beating so fast that it burned raw inside of his breast.

The other cops froze for the sake of desperation and Gyasi rose unnoticed. He aimed and fired. *Swoosh*!

The sound of the liquid spray was audible, but there was no sign of anything coming from the nozzle.

The Iron Soldier took a stagger, as a man stabbed with a sudden pain. It stood in place trying to move but could not. Finally, it doubled over and fell to the ground with a mighty metallic explosion. They watched it anxiously. Its head turned toward Gyasy, its eyes flared threateningly to intense infrared, and then the lights went out of them.

Gyasi had pulled it off! He turned to the cops. They were watching him dumbstruck! "Well gentlemen!" he gloated, "That was too easy! It is not the size of the gun, but the effect of the bullet!"

Halloway walked up to him, carelessly dropping his empty gun to the turf. He was dog-tired! "Well I'll be damned! Who are you?"

"My name is Aduna Gyasi."

"'*The*' Aduna Gyasi! You work for Tutu Babe?"

"I made some stuff for his cooperation," the scientist corrected awkwardly, "but…not that thing over there!"

The cop grinned understandingly, "So I've heard!" He pointed to Gyasi's weapon, "What's that thing?"

"It is a Ray Gun."

"*What?*"

"It is a gun that puts stuff inside of stuff by molecular transmission. Actually, it is science but you can call it *magic*!"

"You're right! I'm too dumb to figure that out! We're all metal heads! But…tell me, what in blazes was in it?"

"Urine."

"Piss? You're kidding me! A gun that uses piss for bullets?"

"Not at all! It is a water gun. We did not have pure water nearby. Look at that big Iron Soldier. A little fluid in its circuit works better than bullets!"

"Yeah! We all got our Achilles heels!" The cop allowed his weary body to fall to the ground. Then he cuddled his own gun and said, "Awesome shit is what you do old man!"

10

May 20, 2020.

Tutu Babe woke with a start. Cold beads of sweat trickled down his face. It was happening again – that seething lust! He sat up in the bed shivering. Then his eyes rested on the blond haired woman beside him. He watched her whilst willing himself to stay calm.

Suddenly the big man put forth his huge bear palm and clobbered her across her face. "*Get up!*" he roared. Her body shot upright instantaneously. He watched her, half-amused. She was grimacing in pain and blinking in utter terror mixed with bemusement. He had struck her so hard that already the blood was flowing from the side of her broken lips. Tutu Babe saw the questioning look upon her face, even as her attempts to scream failed. He obliged her with the answer, impatient as ever. "Get go out of my house bitch! Look at me! You think I don't know you don't like me?"

"*What?*"

He put forth his swarthy hand and shoved her so hard that she fell to the ground. "Ugh!" she grunted as she landed on her back, knocking the air from her lungs and cracking her head against the expensive wooden floor.

"*Bitch!*" he yelled simultaneously through clenched teeth, "Stop pretending! Women only come here for my money – hoe! I'm too fat for you! Up and away!"

She saw evil in his eyes and feared that she would die that night! He was slowly rising to deal with her on the floor, because she was taking too long to go. He needed her out, *urgently*!

"Oh *god*!" the terrified woman lamented hoarsely. "I'm going! Please don't hit me again Mr. Babe!"

She sprang swiftly to her feet fumbling for clothes. The slim built woman glanced across at him and decided she was out of time, so she hustled through the door butt naked. Babe was breathing even heavier now, desperate to quench his lust! He wanted to feed on any available fear and he needed to consume flesh! *Cerebral* flesh!

The energy of her fear quickly surged through his being and dissipated. He heard her bare feet going down the steps, fading away. He needed relief fast!

"Albert!" he called. The drool of his mouth flew across the floor as he screamed. "Fucking come in here now!"

The middle-aged butler pushed the door within seconds and appeared before the troubled boss. "Sir?"

"I thirst badly! I had to get rid of that useless woman!"

"What can I do sir? Should I call Dr. Carter about your addiction?"

"No! I'll get to him later! Go get the little black girl!"

"Yes Sir!" Albert replied urgently and took off for the cellar.

Tutu fell back on the bed and hunched up in desperation. How long would Albert take? When he heard the door come ajar again, he pulled himself up. The girl was there! Albert backed off, pulled the door open even with his back turned to it. Then he slipped through and disappeared, closing the little victim inside.

The eleven year old trembled with uncontrollable fear! Tutu Babe could feel her heart beating – literally, and it was exhilarating to him. That was all he needed! He licked his tongue and smiled at her. "Hi little black girl!" he greeted in a chilly distant voice, "You better be afraid!"

The child did not answer. Albert had unfastened the shackles on her, as Babe would want it. The big man was enjoying this one. Fear was good! Adrenaline was the best invention on earth!

"Little black girl, you don't have to answer me, because I hate you! I hate little black girls and their parents that look for them and can't find them!"

The child began to shed silent tears and that was good reward for his doings. He breathed in deeply. "Guess what now! You better run! I'm going to eat the dead sixth sense out of your fucking head!"

"Eeeh!" she screeched and took off for the door. She grabbed the lock and twisted. Albert had bolted it outside! "Eeeh!" she screamed again, running past Babe back into the room, desperate to find the impossible exit port! She was running, but he was walking all the time. There was no greater bliss than what he felt from her torment! Immortality had its price but he really did not mind the sacrifice! As Carter James told him, the passion for murder and that of sex were close cousins!

When she exhausted herself, she fell on the floor. He grabbed her by the kinky hair and drew her up to his bed. Then he bit her in the face, just to hear her scream to savor it!

"This is sweet!" he told her lustfully. "I must find a way to keep my supply of youth coming!"

....

Three years later, Carter James introduced Simon Muir to the East Lab. He was showing the newbie his latest creation…

"What the heck is that thing Carter?"

"Meat Simon! Don't you know *meat*? Never mind that it's a plant too!"

Simon leaned forward, scrutinizing the specimen in astonishment. He put forth an index finger and Carter shifted instinctively, preparing to stop him from contaminating the masterpiece. Simon withdrew with revulsion, bringing hand and finger back to his side. "I know what meat is idiot! Just that yours isn't normal. Gives me the creeps!"

Carter closed the lid. "Why, because it puffs up and throbs like a hot vagina?"

"*Grubby*! It gets me even worse when you describe it!"

"Do you mean it turns you on mate? I'll punch a bloody hole in it for you!"

"Hell no, you sick Englishman! You and your *meat* make my skin crawl!"

"Aw, it's your dinner you're talking about! Be a man! Do not let a little 'muscle move' scare you!"

"Let the people see this and they're not going to freaking eat it!"

"Don't bet on that theory! People will eat anything as long as it is 'meat'! Americans eat dough-improver made from human hair – every day! That is next to eating people! This is a bloody plant! Tell me which is worse?" He eyed his new assistant mischievously and added, "Besides, I heard you ate monkey brain for Christmas."

"*Dude*, I said I didn't eat it!"

"Oh, you didn't eat it. You bloody carnivorous devil you!"

Simon glared at his friend, shook his head in disgust and changed the topic. "So, what's that device over there for?"

"What do you mean?"

"Above your head Carter – The ticking thing."

"You mean the clock?"

"The *clock*?"

"Sure metal-head! It is the bloody clock – or, that is what we call it. It makes our meat throb like that as it circulates the blood through the veins."

"It's a heart…I mean a circulatory system of sorts?"

"Yes it's a heart then mate! It supplies this entire line with oxygen and minerals."

Simon feigned befuddlement. He patted Carter's left shoulder and joked, "Like yeah, but how on earth does this meat make blood if it doesn't eat?"

Carter grinned. "We have a different type of meat in each line. Look at this specimen and tell me what it is. What do you think?"

"Easy dude! That's beef!"

"Right! There's chicken over there...mutton – We even have fish!"

"*Fish?* You're kidding me?"

"Different kinds of fish too!"

Simon was impressed! His next question would affect his outlook on life. He believed in Christianity and stood resilient in the face of contradicting science. Now Carter was his mentor, who believed in everything but religion! It would take some doing, and much proof, to convince Simon Muir that man had any claim to god-hood. He spoke suggestively, hoping to authorize his own words, "With all we do, we cannot fathom God's fundamental secrets of creation – Can we?"

Carter's grin broadened by a huge margin. Simon looked at him and frowned. "*What*?" the student quizzed with emphasized curiosity.

The lanky biologist took off his glasses and looked down from his heights, into the little man's eyes. "Not because I am wearing these cave man spectacles! Trust me! They are loaded with something other than 'fashion'! If you believe I cannot make and grow meat from scratch, you are a fool mate! I used ready done DNA only as a practical shortcut. With all the advancements in science, we managed to probe into matter too bloody deep to question! I can make human flesh and bone from the dust of the earth! I challenge the imbecile who says otherwise!"

Simon chose to believe that Carter was kidding. Before he could add his own spin to the joke, realization set in. The scientist was dead serious!

"Come on now Carter! You're a mere man – not *God*!"

"Maybe Simon, but I can make living flesh, just like your 'God' did!"

"That's blasphemous! And even if you could make flesh and bone, you couldn't breathe life into it anyway!"

"I could not?" the biological genius asked smartly, testing Simon. Carter was weary of futile debates. He never signed up for a hurly-burly match! All he wanted to do was his job. The defeated subjects never seized! Simon eyed him with uncertainty, hoping he did not have something more to say about it. He was too old to navigate yet another flaw in his religious belief.

Over the last decade or more, masters of Science were ripping Christianity apart. It happened at such a rapid pace that it seemed like delirium. When Simon spoke, he meant it as a direct challenge. Instead, it sounded like a desperate question. "Come on Carter! Creating 'life'? Only God can do that one! It needs no debate! Does it..?"

"You're right mate," the genius conceded, eager to end the monotonous contest. "*I* could not create a living soul in my created flesh!"

Simon relaxed. He had won the argument and his Christianity was intact. "Exactly!" he stressed triumphantly, "That's all I'm saying! We are *great*, but we cannot be like God. Some things we won't ever be able to do!"

"I hear you mate," Carter replied assuredly. He put his cave man spectacles back on and continued, "But creating a living soul will definitely not be one of them! I said *I* couldn't do it, but how about my good friend from college? He's the Total Scientist and master of cyber-genetics, in its true definition. He bloody well can!"

"Bullshit!"

"Actually, his name is Aduna Gyasi – not *'Bullshit'*. You'll meet him when we get to the other building! The people from East Lab have a name for him you wouldn't like!"

"Everyone has heard about Aduna Gyasi! What do they call him?"

"'God' mate!"

"*God?*"

"Word!"

Simon was at a loss for words! How could everything he knew that was right in life become wrong overnight? He eyed the teacher spitefully. The unsightly twentieth century spectacles looked back at him with dark goggles. "What the hell is that ugly thing for anyway?"

"It's an old design with new technology! Aduna Gyasi loaned it to me. It sees people's energy! I can read your mind like 'God' can! Right now, you're agitated, and for no bloody reason at all! Anyway my friend, since you're the first to be pissed off at me, let me be the first to say *'fuck you'*!"

Simon eyed Carter spitefully, shaking his head in disbelief. "Some of the devices you people make in these labs should be banned! I swear!"

"It's another piece from Aduna Gyasi!"

"I guessed that! Did he create the monocle/spectacle view concept as well? All we know is that they come from Babe Coop!"

"Aduna Gyasi invented all the best for the coop!"

A thought struck Simon. Carter realized it and braced for the question. When it came, it was to change the subject completely. "My daughter doesn't want external screen spectacles anymore. She's passed the age and thinks she deserves in-the-head pay per view. If she were to install a contact lens type monocle vision – big screen view, which eye would you recommend for her to put it into?"

"That would depend on whether she's right or left brained. We have an upgraded system. Gyasi sold it to Babe last week. If you wait until September, he's going to roll out the Third Eye View. It'll be a simple implant beneath the skin above the bridge of the nose."

"By golly she'd just love that one!"

At Simon's last word, Carter thought about something. He hated to argue but he had to say it anyway. "Simon?"

"Huh?"

"We have to be the gods! You religious folks are hypocrites mate! Did you see your god creating these technologies? Did he manufacture a fur coat to swaddle your daughter on the way to school in the freezing cold? You'll talk about nature, but you don't want to deal with its fierceness! You want Aduna Gyasi technologies, not mosquitoes on your arse in the jungle! We didn't inherit the best conditions from nature – or 'God'! We didn't inherit the best environment! We inherited the ideal conditioning, which is a superior intellectual faculty – from nature, or 'God'. With it, we had to create and be creating the ideal reality. We had no choice but to be the gods! In all of this, we still haven't seen your god, but winter is freezing and summer is hot! He hasn't objected to you choosing the urban life-style either! What the fuck is the bloody hullabaloo about?"

. . . .

In the early days after the rift between them, Tutu Babe was determined to garner support against the diminutive god-creator. He was on every media channel, bombarding their monocle and spectacle vision. His voice took up the web pages on the internet, compelling all to hear his message.

"I am not perfect, but all the work I ever did was to benefit my country and for humanity on a whole. The Iron Soldier incident was a tragic and unforgettable affair!"

"Although we are grateful to Dr. Gyasi for stopping our Legion, we take into account the fact that he refused

to share his own technologies that would have guaranteed us an incident free awakening of the Iron Soldier. He shirked his responsibilities and we had to seek an alternative. That led to another person becoming our project manager."

"We still seek to console those who lost loved ones on that day. Tutu Babe Cooperation did step forward. We offered to share all of ten percent of the cost, even though it was not our liability. The rogue scientist, Dr. Groh, who deliberately compromised the program, also lost his life in the event. What could we do? Whom could we hold accountable? Surely not the Government of the United States! Let us put it all behind us and look to the future! That is what I need to discuss with the entire world!"

"The future looks bleak! If this message interrupted your program, whether you were using a cell phone, monocle/spectacle or home television, do not change your dial. Please listen to my warning!"

"You do not know the kind of abomination that Aduna Gyasi is preparing as we speak. I know! This abomination is what separates Aduna and Tutu Babe! His proposals are immoral and ungodly! We must respect God in all things! My projects were never to challenge divine authority. They were to work within God's order. Aduna Gyasi is putting forth a different concept that makes man responsible for the divine works of God. If you do not see the difference, it will be a tragedy! It will make the Iron Soldier incident seem like nothing! Remember that this Aduna Gyasy is the same man who caused the death of his entire family and then watched them die without batting an eye-lid!"

"We would create robots – Machines that are not human. These machines are no different from your cars or skaters! These robots would do our bidding as machines. *Aduna Gyasi is proposing to make man into gods, which is a blasphemous concept! What will be worse is this: He will create these beings as a cross between humans and microscopic cyber-systems spliced to their genes and programmed to mutate in awesome ways. They will not come from your seed or mine, but from the testicles of Aduna Gyasi himself! What does that tell you?"*

"Is this a conspiracy to become god on earth? Is it a plot to conquer and shepherd humanity? I am not saying that it is so. The facts give me no reason to believe otherwise. Aduna has always been my friend. He created eighty percent of our greatest products. I have much to lose in parting company with him. If I make such a huge sacrifice for you, let it not be in vain! Let the mantra henceforth become 'the cyber-baby! Kill it in the womb!'"

11

June   18,
2025.

By 3:30 PM, they had formed a one-quarter mile wide circle around the East Lab. It seemed like well over a million protesters, but their numbers were growing all the time. Backup units poured in to bolster the security efforts. Detective Halloway stood at the top of the steps to the main entrance and scanned the area. He turned a quizzical gaze heavenward, to the dark cloudy Boston sky. It was not a peaceful demonstration, but it could have been a hell of a lot worse. Irate citizens followed, wherever Aduna Gyasi went!

He heard another explosion, much larger than all the others before. Dark clouds of smoke rose from Harvard Square.

Anything could happen when the state allows sick scientists to put their abomination labs on the doorsteps of Harvard! The rumors had become real, but the government did not seem to care. The East and West Labs now belonged to Aduna Gyasi. Tutu Babe Cooperation relocated to Cambridge. The big man's propaganda was working perfectly well. It seemed like everybody hated the little scientist. Public opinion did not concern Aduna Gyasi though. He declared his hand. He officially launched the Cyber-Baby Project that year!

All over the earth, it was the hottest debate. The public believed in the billionaire's integrity. Based on Babe's account, they felt that Aduna Gyasi was insane and would be better off behind bars! Unfortunately, the authorities gave him too much room to afflict the earth with what they described as 'his abominations'.

After the god-man separated from Tutu Babe Cooperation, one of his most able disciples, Samuel Uche, defected to work for the billionaire.

There was a bout of intense mass screaming and the young cop cocked his ear. "Heard *that?*" he beckoned with urgency, instinctively heading towards the sound.

"Stay put Charley boy!"

Charley froze and looked at the senior cop puzzled.

"It's OK," Halloway assured him, "We have enough people down there. We don't leave the entrance for anything! If they need help they'll call."

Charley nodded. Halloway was right as always. The young cop's skin raised goose pimples as he listened to the wailing of tormented souls.

"Horrible, isn't it?"

Halloway did not seem perturbed. He stood tall and looked straight ahead, chewing gum like a mechanical being. Finally, after the youngster gave up on a response, "Yuh...but it could be worse, considering the percentages. Most of the population supports the demonstration and we only have this! We should count our lucky stars!"

"Maybe you're right. With all the metal-heads out there though, I wonder what the casualty count is like."

"Me too! Relatively speaking, this mammoth turnout should have overpowered us. The cases of violence have much impact only because of the large crowd. Compare it to some very small demonstrations that are truly violent and it is almost peaceful. The people are not here to make trouble. They only want the political fools to hear them! There are sick-heads in the midst of them, and that's a true reflection of any society in 2025. If they solved the metal-head crisis from around the 1980's, we would not have half the problems today!"

"I know! As the conspiracy theories go, 'shady people' were influencing our government at the time. They figured that heavy metal toxicity would affect the minorities first, making them the 'menaces to society'. Then the less affected majority would agree to race war and mass genocide. That would create the kind of chaos they needed to distract from their dirty works."

"You think someone's influencing our government now?"

"No! That stopped after the early twenties review! Besides, the then minority is the majority now!"

Halloway turned his face down and spat on the ground. "I'm not one for conspiracy hullabaloo! Rats and cockroaches just don't die off!"

Charley watched him in disbelief. "*What*?"

"You heard me!"

"Would you say that if Sam was here?"

"Say what? What the fuck are we talking about? Those motherfuckers been running governments for millenniums and are still there, haunting humanity! What does Sam have to do with it?"

Charley breathed a sigh of relief, especially seeing that Halloway did not realize his lapse in judgment. If the big cop was insulting ethnics, it would be strange.

Halloway seemed mixed himself! It turned out he was anti-something else.

"You're *nihilistic*!" Charley accused in a shocked tone.

"What if I am?" the big cop shot back carelessly.

"Well, how can you be a nihilist and still be a cop?"

"Because it pays me!"

"I know you get paid but how does it make sense?"

"Makes sense because I don't give a fuck! I hate the law, I hate being controlled, but I can be a cop!"

They watched a fire truck speed off toward the sound, followed by an ambulance. That was not the first explosion and it would not be the last. The security forces were winning, but it was a process. Charley would be satisfied with that. There was nothing to get excited about now. In his mind, most of the injured were violent protesters that got what was coming to them. Some were mere opportunists.

Earlier that morning, before they relieved the men on guard at the door, Charley helped with crowd control. They arrested more than a handful of gang members for similar incidents. Occasionally, they had to use reasonable force.

A man was trying to breach the security line, loudspeaker in hand. He pushed for the East Lab entrance, but it was futile. The fool gave up and decided to stay where he was. He turned his back to the security barrier, ready to give his speech. Halloway had seen at least a dozen nuts like him give speeches amidst all the violent chaos since the protest began hours earlier. The person put the loudspeaker to his lips. *"My fellow citizens! May I have your attention please?"* He paused to make sure that the masses were listening before he continued.

"Cyber-managers are encroaching! They already took our best jobs but that was just to begin with. Corporate greed is a disease that we must find a way to cure! Every day they lay off thousands of our citizens. Last week three major companies gave the boot to all their managers and supervisors across the nation. Before, they employed cheap security people to watch and log, while computers performed all managerial tasks. Now even these security people are going home. The machines have become so advanced that they do not need human assistance anymore. They see, they discern, they watch, and they log. They are ever-present, in every corner of the work place. You cannot fart without them knowing! They identify problem workers and take corrective measures. They are tyrants, because they have no discretion – which is immoral – but it suits the stakeholders' fine! Remember the Iron Soldier. These are covert prototypes of the same!"

"Cyber-manager computers are setup as networks in these companies. They open and close doors; they give approvals and resolve issues using their databases. If a solution is not in their databases, they automatically contact a live manager, who could be a company in-house consultant, or a stakeholder on a golf course half way around the world. Yes, they are efficient to the stakeholder, but remember, they are mere machines, without emotions, or any intimate understanding of good or evil. They do not have consciences, because they are soulless entities! That is bad for the worker!"

"This takeover cannot be fair to the people who made these companies prosper in the first place! Next,

they will make all job positions obsolete! I say, away with cyber-managers! Get rid of those computers before they get rid of us! Stop Aduna Gyasi and the Lab in their tracks!"

"Here's Sam and Alvin!"

Halloway's words cut into Charley's thoughts. He turned to see the senior officer grinning at him, "Its Sam and Alvin. They're going to give us a break. Are you planning to join that demonstration kid?"

Charley chuckled, "No! But I must admit the guy is making sense!"

"Yuh! Our jobs might soon be on the line too! Aduna Gyasi claims he only makes robots to help and not destroy us! The results prove otherwise!"

"Devil changed America in a few years! Computers are taking over! They had one working in Lynn. Whenever someone farted in the office, even when no one could tell, it went to Lysol spray the air behind the worker's ass! Remember the Iron Soldier?"

"Yup! *That* was some crazy shit!"

"Doesn't the idea of another one scare you?"

"Naw! I really don't give a fuck. What is to be will be!"

"Well, I'm scared anyway. I heard Aduna Gyasi was against putting his chip into a mechanical being. That's why it didn't work."

"He said it was anarchy. Dr. Babe hated his guts for it!"

"Doctor Babe is a pile of shit himself! How can you create something for free will when it does not have a conscience?"

"He didn't create it with free will Charley! That's propaganda! Aduna Gyasi warned that the Iron Soldier program was flawed. He insisted that it would malfunction and act on its own. It happened exactly as he said it would!"

"It was a mess! Maybe they'll make artificial consciences next time!" Charley quipped sarcastically.

"Do you think the criminals that make our lives miserable don't have consciences?"

Charley shrugged. The two cops walked up to them at the top of the stairs. Alvin nodded and Sam said "Waddup players?"

"Too much to tell!" Halloway offered. "You see the crowd!"

"Yah man! I see."

"It's all about one guy! All the time!"

"Sure!" Sam confirmed, "That raas Aduna Gyasi! And we come to protect him! Alvin had to shoot a kid this mornin'!"

Charley heard the boys talking about the incident earlier. A cop had to do what he had to do. They only needed to detoxify some of those kids. Up to 2020, the authorities ignorantly sentenced many youths to death, although the society was responsible for their predicaments. That morning, Alvin had no choice but to put down the metal head. Had the government armed the security forces with non-lethal weapons, they would save that kid! Aduna Gyasi proposed options that would carry the same range as any modern-day rifle. These weapons would make the subject unconscious for up to three hours. That would be better than making them dead! He saw how distraught Alvin looked and wondered what to say. Halloway was studying him. Some things you will learn only by experience. Silence would be fitting. Halloway dragged the youngster by the shoulder, "Come on kid. Let's go grab some coffee around the corner."

Sheryl started to prepare both their orders as soon as she saw them coming.

"You guys want donuts with your coffees?"

"Yes, please," Charley confirmed. Halloway made that answer serve for both. As they took their seats at a corner table, their eyes followed Sheryl's. She watched the television from across the room. Some biologist, *Carter James*, was on in an interview.

"So, Dr. James, what can you say to convince us that your 'meat' is safe?"

The tall lanky man had to bend sideways to talk into the reporter's microphone, but for some reason, it seemed she was watching the camera instead of paying attention to her subject. *"Well, Marilyn, there's so much to say. I do not even know where to start. I remember back in 2010; some people suggested that the world population should feed on insects as an alternative! It was a sickening thought though! Today, from the walls of the West Lab, I unveil the solution! The scientific community is excited about the positive effects it will have on our civilization. It will cut the incidents of meat borne diseases – every single one – drastically."*

"As we speak, we are in the middle of a Mad Cow's Disease epidemic. Most diseases caused by eating infected animals are on the increase, and we have come up with the perfect fix. Of course, if you do not handle planted meat properly – just like any other meat – it may become contaminated. However, we are giving you disease free, healthier, meat to begin with. And guess what! No one will have to slaughter it to feed you! That should take a load off the human conscience! As you know, by the wisdom of the great Aduna Gyasi, who I admire greatly, human development at a higher spiritual level will depend on a clear conscience for all. His new discoveries in mind science and how mind affects our realities shows that our minds automatically creates our conditions, as individuals, according to what our own consciences declare that we deserve. This solves for us this age-old enigma that some were born to suffer! No! It is merely the results of what we do to our consciences. It is the secret of Heaven or Hell, God and the Judgment! Can we not create purer consciences if we stop tormenting the animals too?"

"Well I'm sure we can! I did hear Mr. Gyasi's speech about 'washing your conscience and creating your heaven'. Amazing! The wonders of science!"

"It is amazing Marilyn! Do you know that livestock farming causes more carbon gas emission than all the motor cars on earth?"

"Get out of here!"

"Well, it's true. Now, imagine what planted meat is going to do for the environment."

"Are you saying that it doesn't emit carbon gases?"

"It does, but it's negligible by any standard. Besides, slaughtering animals is brutal. Humankind has to rise above all this bloodshed."

"Well, Sir, I didn't think about it like that...but yes – you are right! Killing is truly brutal. I always wondered how people ate what they watched being killed!"

"There you have it! Now we do not have to kill anything. It's nothing more than a plant, but it's the same meat."

"I repeat – Amazing! So, when will your 'modern meat' hit the supermarkets?"

Sheryl switched the channel, "Motherfuckin' serpent! How're you gonna go and make *meat plant*? That's an abomination!"

Halloway chuckled. "Nothing surprises me anymore Sheryl. And that stupid reporter isn't intent on asking hard questions! It's like she's on a promotional drive!"

Charley took a sip of coffee and smiled, "It's a paid advertisement Halloway! Stop hating the little cute lady. She's at the press conference now. Tell her how you feel! As for 'plant meat' we'll all get used to it!"

"Not me!" Sheryl retorted, "If I have to, I will chop off a chicken's head for dinner. If I kill it myself at least I'll know that it's real meat!"

"Amen to that! And my conscience will be clear too!"

"Come on you two!" the young cop teased, "These scientists are the true caretakers of the universe and Perpetual Motion – not us. We must 'believe' and not support divergent idiosyncrasies of our own. Get rid of your cave man mentalities, you nihilistic animal butchering psychopaths!"

"Shut the fuck up college boy!" Halloway barked with mocked revulsion.

"Yeah! Shut up!" Sheryl chipped in, supporting Halloway. "These scientists are metal-headed lunatics! Take that Aduna Gyasi for example. He's one insane bastard who killed his five kids and a wife with 'experiments'! Now he wants to do us too…talking about 'conscience'!"

Halloway frowned, recalling the incident. "Yuh! I was a kid then. Didn't that twin daughter live for a couple of weeks? I remember they called her..."

"Anne Eve Gyasi. Her sibling died at birth. She died weeks later and they cremated her somewhere in Spring-Field."

....

While the rioters raged outside, calling for his head, peace and tranquility abode inside the pressroom at the East Lab. The man of the moment stepped to the microphone with his happy face on. He was not going to make ignorant people stop him from doing what he knew he was born to do! Aduna went straight into his speech. Introductions, to him, were a waste of useful time. He was not there to debate, but educate.

"Mankind is what we are from the beginning. It is the original prototype, which, paradoxically, is not the same as the original of eons ago. The first Men of the earth, while being our progenitors, are no longer the primary model that defines what we are or should be. The original prototype is what prevails after mutative

ages of miscegenation – as like now – when we are neither regressing nor advancing. Look into the mirror and see. This is the original of what we are! On this foundation, we will build to become greater beings. We will not change the equations of Perpetual Motion. Neither will we challenge its standards. We will merely continue the cycles that authorize the events of how things are and how they will become."

"If we must meet the ideal standard, we must be driven to that achievement. Humanity is the system that drives us. The Human is Man's worst enemy! He cannot survive forever in his own generic structure. He must become Mankind. He is the body snatcher. The Enemy that leeches on you, as deep as your souls; that determines to destroy you but ends up becoming the source of your perfection; the enemy that comes to replace you, but ends up becoming you. Like the child of a rapist to its beloved mother. That child is the rapist, but it is also what she is. Every ounce of blood is hers, but it is the blood of anarchy too. This is one of the mysteries of Creation."

"I chose this topic at the time of a new cross-road. We can no longer consider us 'humans'. As our evolutions are completed, we are stagnant to the cycles. Stagnation in the universe is death. We look back at history without knowing that we are unlike those before us. They were inferior! We have become the original prototype. We are 'Mankind' by a higher standard of definition. However, we have a serious problem! We have stopped progressing to the ideal, and our civilization is imperfect! We must find the new hybrid to advance us, for evolutionary measures have shut down in our genes. We have out-mutated our latest state of humanity. We need new humans now!"

"So here it is! I know! The questions that remain in the zealous stalwart of ignorance! Must we explain

'God' in perspective every time we come to the platform? Unbelievably 'yes'! – In 2025!"

"Our fathers are older folks we look up to, who in their days looked up to those that were before them. When they are no longer with us, we become the fathers, and the new generations look up to us. Our gods are older civilizations that we look up to, that looked up to higher advanced civilizations in our galaxy in their time. When they are no longer with us, we become the gods that other humans will imitate in the universe. It is a perpetual cycle, and at every stage, we do what the gods before us do and play our parts in Perpetual Motion. At every stage, we improve on the product, so that the future generations will get closer to the age of the Light-Beings."

"Humanity is a state of advancing prototypes. Not even the gods are perfect. Our kind replaced a replacement and became the original. Our genesis code of triple six was the highest advancement to the ancients at that time. Now our replacement is about to step into the arena! Humankind is about to advance to a higher level than most of us could ever dream! This is the year of the human rebirth! The world waits with eager anticipation. It is the birth of our next evolutionary catalyst in our cyber-baby. With higher intelligence and stronger body – We will welcome the new prototype, code triple seven! Where man's number is six, there are two sexes. Where man's number becomes seven, there is but one sex. These children are androgynous and independent! However, their first androgynous soul shall not be born until after seventy-two generations. After this point, there will be stagnation and a need for a new prototype."

"Question here!"

"You! Go ahead!"

"Does *'being replaced'* mean to be made extinct? I don't get it, and those thousands of people rioting outside for your scalp are not favorites to get it either!"

"OK ma'am. You are making me repeat myself here, but I feel your passion. You need to review what I said before. The answer to your question is paradoxical. It is not always easy for the masses to separate the issues and see the difference. I hope you can. You are the reporter. You will have to explain it to them. If the Lab were thinking about making you extinct, it sure would mean wiping my family out and me too. But, see, I don't want my seed to go extinct. The prototype will replace us only by becoming us. What do I mean by this? We will become the same by inter-breeding for seventy-two generations. This is how we became what we are, and how we will become what we will be. In this season, it is the power of intercourse over petty prejudices. After millenniums, it will be the power of individualism above dependency. The first will be the last. The goats will outnumber the sheep!"

"Look on top of your Gothic church buildings. Maybe you will see the image of a rooster – A cock 'crowing'. Maybe you will see a pointed structure raised heavenward. It is symbolic of the phallus. On their windows, that v-shaped design represents the image of the vagina. The phallus and the vagina prevailed at every turn, and so it will forever be. Male and female gender must prevail. There is only one state higher than this. The merger between cock and hen will create a higher being perpetually until they become androgynous! Bitter will create nothing in partnership with bitterness. We are not compatible to 'sameness'. That is why the 'race' game ever fails! Our genes love to be fortified with alien values."

"The cyber-baby will affect a very strong genetic structure in the first new human hybrid, 'Arpago'. As I

explained before, its DNA will be highly programmed and almost indestructible to Perpetual Motion. It will mutate according to internal 'god-codes', designed to change according to the celestial waltz of the cosmic. It will evolve over time to become even better. These hybrids will survive through the bodies of original prototypes by activating effective genes in these inferiors. There will be illusions and questions of race. The cyber-baby of itself will not be the original hybrid. It is the begotten of a god-creator. The hybrid will be its offspring, the son or daughter of Man. In ancient times, its birth was shrouded in mystery as 'immaculate conception', because its 'first-father' is never revealed. It comes from a human that depends on no other but the divine. It defines the best of earth and heaven, brain and mind, body and spirit. It is self-sufficient."

"When the angelic child of us gods impregnate the daughter of Mankind, the offspring will be one hundred – not fifty – but one hundred percent triple sevens hybrid. His programmed DNA will program the DNA structure inside of her. The outcome will be as we, by the power vested in us, have predetermined."

"Who will sleep with such monsters?" The shocked reporter quipped wide-eyed, and then she paused to regain her professional 'swag', "Sorry...that was a shocking thought. But...let me ask you...the cyber *thing*...will you not make any women?"

"No. We are only going to make males. That is all we need anyway. Besides, a female prototype will not work – but let us not get into that one! Sorry gentle men! This insolent woman before me is ecstatic! On behalf of the gods, we are making sons for the daughters of men. Ladies, you can scream with delight now! That includes you ma'am!"

His eyes caught Marilyn at the back of the crowd, raising her hand desperately for her news team. *"Yes, Miss Trooper,"* he offered politely, *"Speak now – I am listening!"*

"Dr. Gyasi...before we all scream with delight, let me remind you – just for the records that you have no family. You said *'it would mean wiping out my family and me too'*, which would seem irrelevant. You claim that you can create living humans from scratch. How is that possible? I know we can manufacture living flesh in planted meat. That is not like creating consciousness, or spirit, which makes us alive and aware. Wouldn't that be a different ball game?"

"That is a different topic altogether! You are a wise young one, so I will indulge you! Your question demonstrates confusion. If you were exposed to religion, you might not believe the possibilities either. Even when their results are right in front of you. Do not apply preconceived notions of norm to reason. Just look for the facts and prove them without restricting your mind. It is painstaking to explain these things. No, I cannot create the spirit being – I cannot create any spirit at all. All spirits exists everywhere. Spirits are inherent universal energies or determinations – just like mathematical equations. You do not create them. You accommodate or use them. They are all the same, ideas, cognizance...definitions. There is no such thing as a single human spirit – There are legions of conceptions – or spirits – in a soul, and all is perpetual energy. If you had only one spirit, you could not even be aware of yourself. The legions intercourse and together they conceive a persona, which identifies itself as 'I am'. This is the birth of a soul. I can create a being by utilizing the spirits that are already available in the universe. These same vibrations are the spirits in electronics. They are the computer's intelligence. They are vibrations of sound and light."

"My cyber-baby will be able to conceive the vibrations of energies – which also are 'thoughts' that emit from the minds of humans. In other words, he will be able to read your mind over a certain distance. Therefore, you could talk to him without spoken words – As you do at times when you pray silently to 'God'. This also reflects what our posterity will become and what a home computer, connected to the web – its universe, is."

"We have advanced to a higher stage in Perpetual Motion. Words themselves are real defining spirits. We need to redefine many of our words to improve on our expressions and intelligence. For example, a demon is an adverse cultivated mindset, idea or determination. This makes it a spiritual, or energy, organization. In a computer, it is a virus. You can be possessed by an evil thought – and a demonic legion, or virus, may program you and determine your outcome."

"I would not need this formula to incite a soul in the cyber-baby, because it will be the child of living beings. I can cultivate a mindset in a physical brain matter by inserting a cult, greater than I in intelligence and faculties. What is a cult? It is a cultivated mindset to the degree of a persona. A demon? Only if it is adverse. An adverse culture makes a demonic nation. We will create higher standards based on true divine principles. We will make humans that look like us and inform them according to our knowledge base. With free will and superior faculties, they will question all things. The cyber-baby's offspring, Arpago, will have a conscience, just like us. He will be a better person."

"I cannot explain these things much deeper, but let our smart children think on them for future generations. If you could catch a photon from electrodes in your brain, you could view a spirit from the legion of your soul...and you are an 'Angel'. Know yourself Miss Trooper!"

The scientist watched Marilyn lean her head sideways to reckon and it amused him. She had too many questions to ask, but he had to move on. She turned to the cameraman in exasperation, "What is he telling me? He does that *every time!*"

"He's picking on you as usual!"

"Dr. Gyasi!"

Aduna nodded to a hairy man in the back.

"Just when did you realize that you could create humans?"

"Let me re-emphasize! The Cyber-Baby Project is not to create, but to modify. We have the faculty and formula to create humans from scratch but that is not even necessary. Aduna Gyasi will not go there! Now, let me answer the irrelevant question. Many years ago, I accidentally stumbled upon the Genesis Code."

"Code Triple Six?"

"Yes. I have worked on it for years. It has been one of God's best kept secrets, and..."

"Triple six is the mark of the beast Dr. Gyasi!"

"No! Triple six is grossly misunderstood. Humans love to play, but they hate to think! When you grab your two sacks of balls and a gulf club in the morning, take reckoning with you! Triple six is your genetic program, created eons ago. It is the code for humanity and for our incubation system of principality, under Angels and Watchers. The human is a mere machine, made into a beast to spur Mankind on to evolution! If 666 is the mark of the devil, the devil is in your blood!"

"That statement is blasphemous Sir!"

"Must that bother me? The number for man, by your own religious calculations, is 6. Who is this god you serve?"

"He is the God of Israel!"

"I know who you mean! Look here!"

Gyasi took up the chalk and wrote on the board behind him. You could hear a pin drop when he was finished! Of all the nations of Christians on earth, no one ever revealed it before! In Gyasi's bold writing, were the following gem atria calculations under the heading 'DIVINE SECRETS OF THE WATCHERS':

God   of   Israel is
42+90+24+90+36+54+114+108+6+30+72 = **666**

God   of   Zion is
42+90+24+90+36+156+54+90+84 = **666**

Humanity is 48, 126, 84, 6, 78, 54, 120, 150
= **666**

Computer is 18, 90, 78, 96, 126, 120, 30, 108
= **666**= *World Wide Web (WWW)*

Witchcraft is 138, 54, 120, 18, 48, 18, 108, 6, 36, 120 = **666** = *'Science Magic'.*

Gyasi eyed the crowd mischievously and said, *"The Law is God. It is every creating, perpetuating and existing equation. A law, when broken, brings destruction. Religious folks belief in ignorance. You define scientific truths as esoteric marvels. The evil you*

*condemn you perpetuate. It is close to your face. You
cannot see it with your eyes."*

Somewhere in the room, one reporter
muttered, *"Jesus!"* while others seemed at a loss for
words. Aduna Gyasi waited, ready to answer the
questions he incited. He would reveal the mystery of a
special gene found in some humans. Then the shots rang
out!

"Die beast! God condemns you to Hell!"

The members of the press scattered for
cover. Aduna Gyasi was too shocked to move! He put
his right hand to his stomach and then looked at it in
disbelief. He looked up and saw the man standing before
him. The culprit pumped off two more shots before a
hail of bullets took him out. Belatedly, a security
personnel dove to push the injured man out of harm's
way. The scrawny shooter in shocking blue was on the
floor, scrambling to get back to his gun. "Kill the
serpent! Save humanity!" he chanted like a mechanical
being.

"Stop you!" Max hollered at the shooter before
firing. The zealot's head exploded like a melon, sending
bloody vapors and minced cranium a good distance
across the floor. Max lowered his gun and winced at his
own handiwork, feeling like throwing up. He watched
the body convulsing, kicking against the odds to no
avail. Every bit of his brain was missing, and all the cop
could see for a face was a scattered row of bottom teeth
hooked on to the remnant of a jawbone.

Halloway came storming in with Charley at his
heels, "What happened?"

"Holy fuck – Don't know! He shot Dr. Gyasi! We thought he was a reporter! Took us by surprise! We took too long to react!"

Max shivered even as he spoke. He was a wiry ex-paper-pusher who was not ready for that kind of situation. Everyone except Halloway believed otherwise. Max could not fit his clothes properly but he was supposed to deal with life and death details!

"It's awright Max. You took him out!"

Halloway pushed by the people that were trying to resuscitate the serpent and went to the shooter. He turned around nodding his head, "You're right Max! He was a reporter from Boston's Day. Face off or not, I knew him! Thought he was nice and quiet too!"

Marilyn stood meters away combating the urge to throw up. She overheard Halloway identify the shooter and the prospect of good news worked as an antidote. The little woman hurried over to him, microphone in hand, "Detective Halloway!"

"Were you here when it happened?" the detective snapped coldly, cutting her off. His voice sounded like evil thunder. She spent all her mental and emotional reserves to deal with the bloody experience. Now her spirit was low and she was vulnerable.

"Y-Yes D-Detective..." The fearless Marilyn Trooper stuttered uncharacteristically. Halloway was a big muscular unfeeling devil-eyed brute with a gun! He was phenomenally cold and he pit-bull barked when he talked! She avoided his arrogant bloodshot anaconda eyes. *Fucking metal head psycho!*

Marilyn heard about him the day she came to Massachusetts. She was eager to meet him. Now they met! Unfortunately, things did not go according to expectations.

"Well," he stipulated, cutting her off again, "you are witness to a crime and I have many questions to ask you! While you scavenge for news, I have work to do! You must leave me alone to concentrate! Please!"

He spun on his heel and walked off, although he said he had to question her. Halloway would not look good on the news later on, but everyone knew he did not care! She would seem like a loser herself!

Marilyn glanced around, studying the frenzy in the room. All of a sudden, the area was crawling with security personnel, from the locals to FBI agents! In her mind, Aduna Gyasi would most likely be dead anyway! She focused on Halloway's broad gorilla's back, and her charcoal eyes filled up with hot-tempered malice. "*Asshole!*" she muttered underneath her breath. He caught something, stopped and half-turned, "*What?*"

"Nothing...Not *you!*" she lied like a coward, the taste of her own hypocrisy killing her. Today was not her day! She would leave him for tomorrow! Marilyn felt abused as she watched him go. If only she could hit him on top of his head with her microphone! Charley froze to the spot in front of her, an apologetic look on his face. All it did was make her more uncomfortable. She turned away and he went to catch up with his partner.

"Marilyn?" The camera operator waved at her face as if to check if she was awake. He turned off the camera and eyed her concerned.

"I'm alright Eric! Just don't say a fucking word now...*please!*"

He rolled his eyes at her swearing. When that '*f*' word parted company with Marilyn's lips, it meant she was livid! "Guy's an asshole! A lead-head or something worse!"

"Some days will suck Eric! I can deal with scumbags!"

"Maybe the problem was not the scumbag. It was that shooting experience!"

Halloway underestimated her! He would show respect the next time they met! She had never experienced a shooting before! Now she was stronger. In the future, she would be prepared!

Charley caught up with Halloway heading to the conference room exit. "Wow! What was that brand of animosity for, partner?"

"I don't know!" Halloway afforded reluctantly.

"You don't know?"

"Are you deaf?" he snapped, "I said I don't know! Maybe I just don't like her!"

Charley was trying to make sense of what his partner said. On realizing that it made none whatsoever, he chuckled. Then he stated matter-of-factually, "This woman has enough problems already...and now *you*, in her face. You should apologize!"

"Reporters don't have feelings! But, if it makes you feel good, kiss her ass for me when she comes around again!"

"You're sick!"

They went back to their post and Charley declared suddenly, "I was reading an article about her last night."

"About who?"

"*Her* – The reporter!"

"Marilyn? You obsessed with her?"

"No, but her story is interesting. She only remembers being in an orphanage as a young child. Then a couple took her in. She knows nothing about her real mother and father. I under..."

"She turned out alright – Didn't she? Let's concentrate on work! The man we're here to protect just got shot!"

"We're only supposed to stay out here and keep the crowd at bay! And by the way, I wouldn't mind kissing that cute little butt!"

"*Ewe*! I'll tell your wife!"

"You'll do no such thing! Stop behaving like she's not hot! You don't have to be defensive!"

"What's that supposed to mean?"

"She's intelligent and beautiful, but she's human! No woman is too good for you!"

12

It was warm in her lonely room that night, but she felt cold any way! This was not her life! It could never be! Something else was out there for her, or she would be living for nothing! Marilyn sat hunched around her computer desk, pulling her warm sweater closer…stretching it around her frame until it threatened to tear. That pullover was her only companion! She should not even be cold, but her internal warmth had dissipated!

At twenty-eight, she was completely alone! Everyone talked about how beautiful she looked, but she never had a boy friend in school. Now it only got worst! Marilyn was always the best at what she did, but somehow it seemed she could not connect with other humans. Was it that she was saying the wrong things at the wrong times? Was it a curse? Maybe, from the parents that dumped her since the day she was born!

Halloway's behavior was like the final push over the edge. She did not know how to blame him. He was just one of many who would not indulge her, although that was a different context all together. Being rude went well overboard!

The one person who seemed affectionate around Marilyn was that devil. He always focused on her. Aduna Gyasi press conferences belonged to the little woman it seemed. Sometimes she heard that still voice in her head, even in her sleep, *'Go ahead Miss Trooper. I am listening'*. If she did not know the serpent, she would think he was fond of her. She always got the best of him, and sometimes it seemed conspicuously biased. Who cared? It only meant notches to her status at the top of the game. She figured it all out! Aduna Gyasi loved challenges. She was the one who challenged him the most, because she was the best at what she did!

Yes, Marilyn Trooper was good as a professional. Every woman wanted to be her. They did not know how she sucked in life!

What problem did this godforsaken world have with her? If she had parents, if she had grown up in a conventional home, would her life be different? Yes! She was sure! Even at her adult age, she needed her mother...the one she never knew! Most of all, she needed a father to run to. Of all her needs, that one gnawed at her insides the most!

If she had a father, she would call him up right there then and he could help her get through seeing a man being shot multiple times. She wondered if she would ever live that one down...even though the victim was the devil!

The television was on. Marilyn cupped her face into her palms and listened to Stefanie Benson's 'All the Children Matter'. The program highlighted people like herself. Stefanie was an orphan. She knew how it felt to face emotional trauma because of absent or delinquent parents.

A man who claimed to be an ex-gangster was talking about his experience. After all his suffering, including seven long years in prison, he got lucky. He tried out for a plant meat franchise, expecting to fail. He was pleasantly surprised when Tutu Babe Coop gave him the call! Then his father appeared from nowhere. He could not forgive the old man, no matter how he tried. He ended by saying, "Damn those people who think they will piss out babies and leave them on door steps!"

Marilyn nodded in agreement. "Damn them!" she confirmed with ice-cold venom.

....

The masses of Massachusetts celebrated jubilee when the first news spread about the shooting. Then the truth came and their joys dissipated.

Gyasi was a master at the East Lab, and his associates were the best at saving lives. Weeks after facing certain death, he was calling another press conference, alive and well, and he had his happy face on. His fatal wounds had miraculously healed, just like that of the beast of Revelations. Now, the believers had no reason to doubt. They officially declared him *'the Antichrist'*!

Aduna Gyasi was always the big news. The media could not afford for him to die! Like any great star, this mighty son rose from the grave and back into orbit.

Zealous fortitude turned patient words of love to hateful actions and peace into all out war. Many stopped waiting on God to act and win the battle for humanity. Some churches lost faith in *turning the other cheek*. They were now fighting to win this world instead of living to go to Heaven. In those short weeks, the Faith Defenders Organization grew rapidly. Mercenaries rose to join the payroll, and no one cared if it was for the love of God or the love of money. Even the greatest of atheist now received merit as *'Christian Soldiers'* defending the principles of religion. The irony did not stop at that. Just as it were at the time of Jesus birth, Herod had returned in the form of the Faith Defenders. Someone had put out an anonymous call to kill all young boys under the age of one! Unborn children and their mothers soon came under bloody and ruthless attack at the hands of the zealots. Anarchy had come to the state of Massachusetts to spread all across the globe. The Law could hardly prevail!

No one knew where Aduna Gyasi was. The demonstrators came as usual, but this time, they could not pick him out, even after 3PM, when the press conference should have begun. Some called it a hoax. Either he was dead, or he was not coming.

Inside, the people from the media were getting anxious. They were told that the Lab would do all that was in its power to make sure their master was safe. With all the security troops outside, there was no reason to fear a riot. The only thing he had to fear was, judging by the last meeting, the press...

Then, at 3:30, he came! He just *came* – appearing in the room from out of nowhere! *"Oops!"* He chirped, his voice echoing from every corner, *"Sorry I am late! Technical difficulties with teleporting!"*

They watched and listened in awe, and the world watched and listened too! Before he could speak again, Marilyn cut him off. "Sir, it's good to see you here! We thought we lost you weeks ago!"

"Thank you my little princess!" the wizard replied graciously, *"I never doubted your love for me!"*

Marilyn flashed him a deliberate 'artificial' smile. "Well, since we've managed to get through the formalities; are you having any second thoughts about your project? Most people think it's a threat to humanity."

Aduna took a couple of seconds to muse over the little woman's unaffectionate behavior before smiling back warmly, *"A threat to humanity? On the contrary, it is a blessing! I will tell you what a threat to humanity is! Mechanical thinking robots are the threat. The creators of these receive more support than I do. If they had access to my genetic code, they would have created destroyers of Mankind long ago! Still it will come!"*

"The sheep mind is dumb! It is not funny anymore! Even a murderer tries to keep his own family safe. An entity that becomes one with you will not threaten you. The ones that are threats will be those with perceivable differences. Like unconscionable cyber monsters among inferior humans!"

"I told Babe and all the other scientists, and I will tell you now – I will tell the world in this! You do not have to make machines from iron to create invincible warriors. When the first prototype that was Arpago came to earth, it was fully human, but fire blazed in its footprints! What do you want to do? Do you want to create and watch great robots destroy you, or do you want to become Arpago? They tell me about every external source of power and power supplies. I do not care for primitive reports! If you make a loop between your index finger and thumb, that loop is infinity, and I can find any resource in the small invisible space in that loop. I dare them to challenge my greatness! Wherever there is a molecule, there is everything! In fact, wherever there is anything, there is everything; and the source of it all is infinite! I am Aduna Gyasi, a divine child of Perpetual Motion, and my works are the works of God in evolution!"

"Mr. Gyasi!" a man chipped in from the back, "I will let the blasphemous 'God' statement pass. Are you aware that the state is going to close down the East Lab and prevent you from creating this cyber-baby monster thing? Remember what happened when you tampered with perpetual design to create your own offspring!"

"Cyber-baby monster thing? He will be human! Great Gyasi's greater son! He is not going to look anything like you, so he will not qualify as a monster – like you! And for the record, I never tampered with perpetual design. I worked with and embraced the laws

of nature. What is to be will be! Stop bugging me about my past!"

"Now, I do not care about what the state will do! My allegiance is not to a single state. It is to humanity! All our work is for the good of humanity!"

"Many of you condemned Dr. James's planted meat. Now, weeks after it hit the supermarkets, you all cannot get enough. Why do you always oppose what we create for your own good? The state will not, and cannot stop this mission. Aduna Gyasi does not work from the old East Lab structure anymore! If they can find me, they can stop me! Now, no more questions from you, 'monster thing man'! You, speak ma'am!"

"Thank you Dr. Gyasi! How are you going to do it? Are you actually creating this being, or..."

"I know what you want to ask – Although you struggle with the words. I am working with a very gifted biologist. He will give me very fine specimens of human sperms, which we shall program into the new advanced Genesis Code."

"So, you will inseminate volunteers – I am guessing."

"No."

The woman scratched her head and asked sarcastically, "How will you do it then? Sell it to a sperm bank?"

"Sorry. Good question, but I have to go! The day will come when our hybrid shall be the savior for humanity. I do not work to gain power! Wait until you see the true abominations! When it is too late, you will

know the difference between good science gods and evil ones! And you Miss Trooper! You better start deciphering!"

Aduna Gyasi disappeared, leaving the press staring in disbelief. He did not afford them more than a whet appetite! He appeared and disappeared. Was he really even there? Could a hologram be so real that it seemed physical?

By 4:30 PM, the Faith Defenders went wild, storming and destroying sperm banks all over the country, just because of one reporter's sarcasm. The debate between the churches rocketed to fever pitch. The majority of Christians stood firm in their non-violent approach while others, led by the Faith Defenders, went their zealous way screaming for the blood of the antichrist.

There were even whispers from insiders, saying that the powerful Tutu Babe incited and funded the Faith Defender's crusade. No one could link the two together, so it passed off as rumors.

That night they held a major meeting in Harvard Square. The crusaders were there. Alfredo was the person given the task to motivate the mighty Christian Soldiers. He took the stand and all the eyes of the faithful rested upon him.

"Fellow Christians and all concerned sons and daughters of earth. Our world is about to be destroyed by satanic forces! We need to act, and act now!"

"Do not be bamboozled into thinking that these things will go away, or that we must wait upon the Lord! God helps us when we help ourselves in his holy name! If we ignore this, we will create the cardinal sin! Think about it. All over the Bible, especially in the Old

Testament, God relied on his faithful to act and destroy anarchy before it could take hold of the state and contaminate the culture of his faithful fold. That is why he ordered Israel to destroy Canaan, including their bastard children, babes and suckling – Every one of them! He did not ask them to 'wait on the Lord'!"

"Today we received a terrible threat, from the mouth of the serpent himself! That demon being named Aduna Gyasi! Will God tell us to sit back and wait this time? I don't think so! We can defuse this threat! We can disarm the beast! We can put down the attack of the devil by the might of the soldiers of the cross! The Beast intends to bring his offspring into the world, by the womb of a whoring woman! Such serious threats call for drastic measures! Our only recourse is to stop it from happening. The Cyber Baby must be destroyed! Let us act now so that, within months, we will not regret it! Let us make sure that the woman who carries this child does not escape our net. Let us suspect every bastard, if it is as much as a year old! Let us kill the fetus from its mother's womb! The cyber-baby – Kill it in the womb! Kill it in the womb! Kill it in the womb!"

It was a massive gathering of boisterous Christians and secular folks alike. Most of them did not like the tone of Alfredo's speech. Instead of unity, there was division between the zealots of faith. Most believers felt that the organizers had ulterior motives. They accused them of trying to makeover Christianity in their own image to fulfill a worldly agenda. Skeptics soon called the Boston Police. That marked the beginning of a rift between Christians. Some rose up to speak against the 'ungodly' intentions of so-called crusaders. Alfredo announced that those who were not for it were enemies too. That pronouncement magnified the tensions! Seeing that Alfredo *'George-Bushed'* the innocent, many people began to leave in fear. An attack against them seemed imminent.

The police came to control the crowd, respectful of Alfredo's freedom to speak. When they heard the message, they promptly broke up the meeting. Still, enough was already said to excite zealous passions in the evil-minded minority. Their bloods boiled with unholy thirst to quench. They would take action! The crusade started in earnest.

Soon the state had to establish 'safe houses' to protect expectant mothers and babies up to one year old. After a while, the Faith Defenders infiltrated the state-run institutions that the government 'safe-houses' were nicknamed 'sitting-duck houses'. No one wanted to go to them anymore!

Next morning the world awoke to an unusual sight! From the western sky, there shun a huge magnificent light! A silver star it seemed…or it could be a planet, as large as this world. The euphoric energy of the object vibrated through the bodies of all, magnetizing at one period of the day and electrifying at the other. Christians believed that it was the New Jerusalem coming down to Earth. By the mouth of Marilyn Trooper, Aduna Gyasi begged to differ as usual. He claimed it was nothing but a plane, like the Watcher's Phoenix. They were observers from higher realms than even the Watchers, drawn to this dimension because of the activities performed by Watchers and the new 'rotations' of cosmic cycles. Whatever it was, most people agreed that it was a sign of the end times!

13

"You think Mom would endorse what we're doing?" Atom questioned speculatively.

"I don't know!" Omar snapped at him, "But it doesn't matter! We have bills up our asses! Alfredo is hooking us up! How else can we repay him? Mom told us to find jobs and this is a 'job'!"

"Sure!" the smaller twin shot back, "We're in the business of committing murder and selling fetuses!"

Omar was pushing the door to get out of the car. He pulled it back shut again. "OK Atom! You win! Do you want us to quit and go home now? I'm not sure that'll pay the loan tomorrow though! Alfredo will kill us!"

Atom swallowed hard. "I was not saying to quit! I was just 'saying'…you know!"

"Sometimes it'll do you good to shut the fuck up and don't *'say'*!"

"*Cool*!"

"*Fine*! Now let's go!"

Atom was about to say something and his brother waved a warning finger to cut him off. "*Uh-uh*! Are you going to *'say'* again?"

"No fool! I was just going to add a little cerebral thinking to the plan."

"*'Cerebral thinking'*? OK, let's hear it then!"

"*Right*! I was going to tell you…we don't call each other by name, no matter what!"

"I thought that would go without *'saying'*!"

"Well I'm just making sure!"

"Awright! Let's go!"

"We need to 'get dressed' first fool!"

They got 'dressed' for the 'occasion'. Then they hopped out of the car, headed through her drive way and up to her front porch. It was after eight in the evening. The twins had adorned themselves in spanking new Faith Defenders robes, with white hoods covering their heads.

They knew this place too well. It was out in West Fields. The house sat on two acres of property. No neighbor would see them approach Miss Simpson's front door.

Atom kept tripping on the skirt of his garment. It was too long for him, and very uncomfortable. It did not help that he was only five foot three either. "We should have tried these shits on before we came out here!" he agitated. Then he side-glanced at Omar and drew back in shock. "What the *fuck*?"

Omar was six-foot three inches tall and pencil thin! The robe he had on was so short that it went well above his knees. Atom could see the legs of his underpants beneath it. Omar looked at his brother wondering what the matter was. Then he looked down at himself. "What the *fuck*?" he reiterated, flabbergasted.

"Makes no sense to me! They give you a short one and then give me a tall one!"

"It is what it is Atom! Too late to fix that problem now!" Omar was looking around Miss Simpson's lawn and his twin was watching him. "I know what you're thinking!"

"No more lawns for us to cut after tonight!"

"After tonight, we'll be free from loan sharks. There'll be other lawns to cut!"

They got to the porch and stood there wondering what to do next. "How do we get in?" Atom asked childishly. Omar pressed the bell and his smaller brother jumped, startled. "You have the knife?" the skinny one asked.

"Yup!" Atom whispered nervously. He drew the huge chef's knife from his waistband and handed it to his brother. "*Here*!"

"Good!"

Omar kept the vegetable knife behind him and waited.

Suddenly the door flew open and Miss Simpson appeared in front of them. "Aagh!" the twins screamed simultaneously. They did not expect her to open it without checking! She was a brave woman, out there all by herself!

The soft-spoken middle-class woman saw the infamous robes and a bewildered expression crossed her face. "What in blazes..?" she uttered sarcastically, quite unafraid, "Is it Halloween again?"

She looked down at her protruding belly, and back at them, disappointed. "*Omar*? *Atom*? Shame on you two!"

Omar felt his face to make sure the hood was fitting properly. "How'd you know it's us?" he asked. Atom nudged him in the back and he rephrased, "I mean, it's not us at all ma'am, but why do you think it's us?"

"Go home!" the woman warned graciously.

"Sorry ma'am!" Omar said, wedging his left foot between the doorposts. "We have to talk to you!"

She saw the knife showing from behind the lanky brother and said, "Go and talk to your mother about your behavior!"

Atom held up his hand in a truce before butting in, "Miss Simpson, it's not what it seems like!"

"What is not what it seems like? This is what you do now? That is *low*!"

"But ma'am, we have bills to pay!"

"I know! All because you are lazy!"

"We cut your lawn!" Omar retorted in defense. He felt Atom's poke in the back. "I mean," he corrected, "We cut lawns!"

"Well it is not enough! Cut more!" She shoved the door, but the lanky twin's foot was keeping it from closing. "Oh come on ma'am!" Omar coaxed, "We come in peace!"

The woman looked at the knife exposing again. "I can see that!" she replied backing away from the door as the men entered the house. "I will just head up to my room now!" she continued, easing up the stairs, "Then I will return to you in peace!"

As she took the stairs, Omar slipped the knife into Atom's hand, "She's going for a weapon in her room! Get her before she gets it!"

The pregnant woman labored off and the little man went in hot pursuit. "Don't let her scream!" the tall brother advised. "We don't want the neighbors to hear!"

Atom rushed into the room after the woman and the door shut behind him. "Good job bro!" Omar encouraged nervously, "Do that thing!" He tried the doorknob. It was locked! He heard a rustling sound inside. Then his brother shouted, "If you raise that right hand I swear I'll cut it off, bitch!"

"Way to go Atom! Gut that bitch!" he voiced supportively at the doorknob. Then he heard a swishing sound, followed by another; then a tinkling, as if the knife fell. Atom screamed "Aagh!"

Omar tensed. "*Atom*?" he questioned hoarsely.

"Aaaaghh!" was the answer!

"*What*?" Omar poked.

"A nan fucking chaka!" Atom informed him.

Omar heard the woman's soft voice, reprimanding, "So two of you will gang up on a pregnant woman? I'll fix you one at a time!" *SWOOOSH*!

"Aaagh!" the little man screamed again.

"You're freaking me out Atom!" Omar advised edgily, "You got a foot long fucking knife! Stop screaming like a bitch!"

"She took the knife Omar! Tell Mama I'm *dead*!"

"Don't say you're dead! Come, let's forget this! Fuck Alfredo and fuck fetuses!"

"She's not letting me fucking go!"

"Try harder! Extricate yourself from that violent situation!"

Atom paced around at the top of the stairs wondering what to do. Finally, he stopped and called out, "Miss Simpson? Could you please send my brother out now?"

"I'll be talking to you as soon as I'm done with him!" she returned casually.

"Omar!" the little man called, traumatized.

"Yeah?"

"She's using martial arts! Call the fucking police!"

Omar shook his head with incredulity, "How're we to call the cops Atom?" He paused and then addressed the woman, "Miss Simpson! It's illegal to hit my brother with a nanchaka! It's not even a fair fight!"

Suddenly, the lock turned and the door flew open! "Your turn!" the woman chirped at Omar. Atom blew past him down the stairs, holding up the end of his garment to avoid tripping. Omar saw his brother's swollen and bloodied face and screamed "Aaagh!"

The woman raised the big chef's knife so he could see it. "Alright! Nanchakas are illegal. Let's use your foot long knife to gut you!"

Omar's eyes bulged in their sockets and he gasped for breath as if having a seizure. She froze for a second too. It seemed like she was affording him timeout! Finally, he caught himself and sped off to catch up with Atom. The end of his robe was flying like a dancing girl's outfit. Miss Simpson grabbed at it. She missed and fell headfirst, tumbling all the way to the bottom.

Atom was outside, fidgeting impatiently for his brother to come so they could go. Omar stopped and looked behind him. Miss Simpson appeared to be dead! The tall man grinned with irony, his scattered teeth protruding from various angles. "Atom, come in here! She just broke her fucking neck coming after us!"

"*What?*"

"Come and see!"

The little man stepped timidly back into the room. Omar walked over her and took up the knife. "What do we do now?" Atom asked unnecessarily.

Omar pointed to her huge stomach. "Gut it!"

"*What?*"

Omar eyed the little man agitated. He handed over the big knife and repeated slowly for him to hear. "*Gut...her!*"

"Gut her?" Atom echoed imprudently. Then he shrank back. "No way! You do it! There's gonna be too much blood! Plus…suppose she's not dead? I took all the blows today! It's your turn!"

"She's fucking dead fool!"

Miss Simpson's body convulsed. "Aagh!" Omar screamed, grabbing Atom for comfort.

"She looks *dangerous*!" Atom surmised. "What do we do now?"

"Give me the phone! I'll call Alfredo! She's a violent woman and I can't deal with this 'kiyaa-hiyaa' shit!"

They heard someone coming! By the time they looked around, cops filled the living room. Atom handed the knife to his brother. Alvin's gun made a beeline to the skinny man's head. Omar dropped the weapon and stiffened. "Hands up in the air idiots!" the police officer barked.

Miss Simpson moaned and rolled over on her side. "My baby is coming!" she warned Sam! "Too bad I can't stick around to kick some more asses!"

…………………..

He was the enemy, so she was reluctant to let him enter. Erotic truce creates the most enchanting affair!

He came closer to her! She was standing beside her bed, feeling shy as ever, and failing to cast her soft voice across the room. "Please don't come in here."

"Huh?" He asked again, his knitted brow showing how hard he had to concentrate just to hear what she was saying. Every time she spoke, he went closer.

"I don't want you in my room."

He was too close now! She felt his breath. It made her fear the impossible, although he could not have the audacity to try! She hated his guts!

Suddenly, she felt his lips brush against hers! He had the nerves! She found herself going backwards, and then realized that she was in his strong arms. He was laying her down! She felt safe...as the mouse that had a lion for protection. Her resistance turned to ecstasy. She admitted it to herself. This was what she longed for!

She held him desperately, feeling his power...kissing him. He made his way inside of her. It felt so good; she trembled uncontrollably, overcome by emotion...then pain! Suddenly, he became too big...and way too rough!

"John fucking Halloway!" she screamed and woke up!

The little woman tried to bring her body upright, but she could not move! Something strange was happening to her! Marilyn was desperately afraid! She kept her eyes closed in a bid to hide from the unknown. Her heart was beating so fast, she was sure that it would explode. Was she going to die? Where was she, and why was she tied up?

"I see you were having sweet dreams about a certain gentleman my daughter."

Marilyn froze, wishing to play dead although it was futile.

"You are safe. I could not hurt you!"

There was sincerity in his voice and that encouraged her. She heard urgent shuffling in the room. Something was happening, which they did not want her to see! Marilyn opened her eyes and looked around hastily. A woman dressed as a nurse hurried through a door and then closed it behind her.

Aduna Gyasi was a short character with a foot long white beard and similar hair that was shoulder length. The wizard did not seem that old for a man who accomplished so much. He beamed down at her affectionately.

"Why Dr. Gyasi? I really didn't expect you to stoop so low!"

Her ridicule affected him. That was encouraging to her!

"Well, thank goodness! At least you asked why."

"So tell me already! How did I get here and why?"

Aduna Gyasi watched her uncertainly. Marilyn waited. He said nothing.

"What time is it and what day?"

"Do not worry. We got you last night. It is only 7:15 AM the next day. "

"Sunday morning?" She asked looking down at herself. She had on her red lingerie!

"You were not sleeping for days like Sleeping Beauty! You dreamed about a certain Prince Charming none-the-less."

"*Oh!*" she cringed, embarrassed. "Another nightmare! When it isn't one evil brute, it's another!"

He leaned his head sideways to think. Then he offered soberly, "He is probably a better person than you think!"

"*Who?*"

"Detective John *fucking* Halloway. But it goes for me too!"

"Oh." In the middle of her embarrassment, terror suddenly set in. "What did you do to me Aduna Gyasi?" She shuffled to free herself. Both her hands and feet were tied to the four bedposts!

"Do not panic Miss Trooper. We would not do anything to you without your permission."

"Did I give you permission to abduct me?"

"Oh no! See, that is different."

"Why?"

"We needed you and we knew you would not come."

"I'm a reporter! I'd come for good news!"

"You would not come Miss Trooper!" Gyasi asserted. Then he told her the premeditated lie. "I need you to record our doings here...for the good of humanity!"

"*For the good of humanity!*" she mocked disgustedly, "Everything is *for the good of humanity*, but only you serpents ever benefit!"

"Marilyn," he pleaded, sounding genuinely hurt. He could literally taste her mockery in his mouth. Gyasi was forced to mix the truth with lies for the sake of telling the truth. "I never had the opportunity to explain myself at length to any reporter. I am begging you to listen! When the truth suffers the people pay the price!"

Marilyn cackled sarcastically, just to show how useless his efforts were...but then she caught his eyes. They were transparent. He was not in the mood to deceive her! A surge of anxiety flooded her being and her heart fluttered desperately. This was Aduna Gyasi! Something huge was coming! Marilyn knew why he would pick her as a reporter. The whole world would guess why too! She was scared and curious at the same time. Would she fall to some deceptive plot? She frowned with mocked skepticism, "Should I believe you Mr. Gyasi?"

"Well," he affirmed, endearing her, "I am hoping you will my daughter, at least for every time that the truth matters most."

She frowned at his reply, leaning her head sideways to figure. Finally, she ended up smiling over it. All she could utter then was "*Oh*?"

"Oh!"

"So you would bait me with a lie to get me to listen to the truth?"

"Oh sure! Why not?"

He watched her, a knowing look on his face. Aduna Gyasi smiled and waited.

"OK," she conceded, "Don't know if I will believe what I hear, but, I will listen. It's good news anyway!"

Gyasi's face lit up and he managed a smart chuckle. "Good for you!" he exclaimed, "You are as stubborn as I am! It intensified the suspense!"

"I'm nothing like you! You have me tied to a freaking bed!"

"I know my daughter – I know!" he agreed, eager to gratify her.

"Will you please stop referring to me as '*My Daughter*'? It's unbelievably agitating! If I were your daughter, I'd hang myself! At least if I was not roped up!"

"Alright Miss Trooper! My bad!" he said and walked off to hide the face of shame. Marilyn almost felt pity for him. He was the one with super thick skin, who said anything to anyone with no regard for his or her feelings! Now, here she was feeling sorry for hurting his...and all in the middle of him having her tied to a bed! This was not the Aduna Gyasi she knew, who could not care less!

The wizard returned, smiling down at her assuredly. "If I were your father Marilyn, you would never have a life! Not with that conventional mind of yours. If I were my father, I would have to be a chip of the old block to survive in this torrid world of conventionalism. You are a cherished reporter. Use your analytical ability to interpret whatever I reveal to you. Ask the question, receive the answer...understand what it is. I have much faith in you. You alone have that potential!"

He raised a file that was in his left hand, rifled through it with his right, pulled out a page and was about to hand it to her. Then he remembered that she was tied up. He beckoned to Weasel. The assistant hurried over and untied her. Gyasi waved him away, "Now excuse us please!"

He turned his attention back to Marilyn. "Stay or go Miss Trooper! It is your call!"

"You promised me good news! I'll hang around to enjoy your hostility!"

"*Hostility*?" he lamented with mocked concern, "You are hurting our feelings here!"

She sat up in the bed and took the sheet of paper from him. "I am in no position to betray your trust," he whispered confidentially.

"How come? Especially after kidnapping me!"

Marilyn read from the page, *"Anne Eve Gyasi...born August 27, 1997..."*

She stopped reading and eyed the wise man suspiciously. "Who's this *Anne* character? Is she your volunteer?"

"Not really. She is my prospective volunteer."

"Why *her*?"

"Because she is 'Eve'. She has something special inside her womb."

"What would that be?"

"My flesh and blood. I do not give anything to others that I would not give to my own! It is more than just sentiments though. She is my trump card! This project is taking a fight from too many quarters! In Eve, the entire project resides. Everything else is unreal. We apply the same level of effort to create the decoy as we apply to the project itself."

Marilyn rolled her eyes heavenward. "*Oh*! You confuse me!"

She was tugging at her shoulder length black hair with both hands, as if to tear them out. Gyasi shrugged with mocked apology. Marilyn sobered up. She eyed the old man cagily. "Where is she now?"

"Here!"

Again, he ruffled through the file and took out an envelope filled with pictures...but she froze before holding on to them. "It's *me*! Isn't it? Tell me so I won't have to face a wicked surprise! Am I going to see *my* baby pictures?"

The elder nodded, confirming her fear and her gut said that he was not lying to her! "You are deeply intuitive my dear!" he offered approvingly.

"And you are a very wicked man! Damn you!" she screamed with raging hysteria. The little woman threw the envelope at Aduna's face and sprang to her feet, intent on landing punches. Instead, she ended up clinging on to him, desperately trying to not fall. Her body was shocked to weakness and she needed support.

Just when she gave up trying to find her parents her dad arrived! And of all the persons in the world, he had to be the antichrist! Was he the mystery benefactor too? That would be more likely than not.

Marilyn eased him off, and he tried anxiously to catch her eyes, for the sake of damage control. As expected, she did not take it well. He was the menace to the society!

"Ewe! Mr. Devil Daddy," she began, cold as ice. He discerned a hint of disillusionment in her voice. "Is there any rope around here? I'm ready to hang myself now!"

Gyasi hated the lack of sarcasm in her statement. Unwittingly, his eyes went to the pieces of rope left on the bed. Weasel appeared from nowhere, took them up and stretched them to her, eager to please. He saw Gyasi's eyes flare and pulled back hastily, "Oops! Sorry! That's an innocent lapse!"

"Get out ill-mannered bastard! This is a private conversation!"

Marilyn turned her back to Aduna, sending him the message. She was a tough girl, and that scared him! A chip of the old block, although she would hate to know it!

Suddenly, a thought came to her mind. She could not resist it, so she turned around to face him. "How come your skin and nails seem 'alien'..?"

"Because I came down through the ages unpolluted," he interjected with full understanding. "Inky's secret!"

"I'm befuddled!" She countered reflexively, "Who'd guess that you'd be caught up in this modern-day pseudo-religious nonsense? From one impossible bible to the next – huh? Now it's this myth about some ancient space devils that claimed nobility on Earth!"

Gyasi smiled. "It is no myth to me. I know my bloodline. There were different versions of hybrids, from before the great deluge and after. I am a full-blooded child, preserved from the lineage of Mighty Inky. God, history and Mother Earth protected my blood for eons. Your mother is from Mumbai. A hybrid of hybrids she is. You look like her in skin, but you look like me in mental faculty."

"You mean she 'was' from Mumbai!" Marilyn corrected, determined to be rude, "There's a difference between past and present!"

"There is no difference Miss Trooper! Besides, your mother is alive!"

Marilyn stiffened and then turned to face him, her eyes blazing with rage. "And my twin brother?"

"Angel is alive."

"'*Angel*'?" she puzzled, "*Know myself*? There was something to your sayings!"

"There are always things that I should tell you...but I cannot..."

"And there are things that I should do to you...but the law wouldn't permit!"

14

After news broke that the star reporter was missing, the Massachusetts police went to her home to get a DNA template. Then they hit the street, scanning the entire state including private homes, with Aduna Gyasi's amazing Body-Locater system. Nothing turned up! The search extended its borders to every state in America!

Faith Defenders immediately sent out a release to say that Aduna Gyasi was responsible for the reporter's disappearance. They made this accusation because Marilyn Trooper worked assiduously to expose the wizard and his evil designs! To them, it was the perfect motive to silence her. They assumed that she would most likely be dead. That last point was almost indisputable to law enforcers.

The Defenders took to the street too, intensifying their attacks against pregnant women and children under one year old. In their minds, they were 'purging' to prevent the incarnation of Lucifer on earth!

That day, terrible things happened under the lights of the new silver sun. By evening, when our own light-bearing sun appeared to pass the alien, blood came as a recompense for the sins of humanity…but it came from the body of the innocent!

It was far more intense in and around Boston. The city was virtually on lock down, that the Mayor had to ask the residence to stay home. From the time of the Faith Defender press release to the evening of the following day, they heard great explosions, sporadic gunfire and awful screaming…but in many instances, they experienced it!

By the end of that spiral, seventy-two toddlers and women, suspected of being pregnant, were massacred in the region of Harvard alone! They left the women to rot in the street, while the toddlers and fetuses disappeared. The Faith Defenders wreaked havoc against the hospitals in Boston. They invaded nine maternity wards on the day, seizing a hundred and thirty two newborn babies that they would 'sacrifice for the cause'!

Mayor Gilmore went on every news channel to plead for sanity. He spoke with confident words, but his demeanor revealed his fear of losing the war. The Federal Government was sending help. He stressed that it was not half enough! Anarchy had taken a hold on the state! It threatened the entire country…and possibly the world!

That Armageddon's day, for the need to survive, the Boston boys answered the call as sworn protectors of the people. Seven police officers were killed…and it took the lives of twenty-four Faith Defenders to justify their deaths.

On Capitol Hill, they convened a meeting between major stakeholders. Haughton Shield, a representative from the FBI proposed, "The best thing to do is to take out Aduna Gyasi. I don't care if he's innocent! The end will justify the means! Besides, he's involved in illegal activities anyway! If I remember well, the US Government did order the Cyber-Baby Project shut down. Well, the problem we face in our country today is because Aduna Gyasi did not relent! Can we sacrifice state control for private notions of moral, Mr. President?"

Both the President's eyebrows scrolled horizontally across his forehead to touch. The one dozen officials turned their attentions to him. His eyebrows returned to their places, and it seemed that he was ready to speak…but then they slid back to that concentrated position. The men waited again. Finally, the great Commander was ready. He turned to the Secretary of Defense. "How would you propose doing this thing to make it seem like it never happened?"

"Mr. President…I…er. You see…the problem here is not how to do it. The problem is…we can't do it if we cannot find this little man!"

"My god! Who the devil is this fellow? We can hold no power over him! How can one little guy be so illusive?"

"How? The answer is 'science'! It is science beyond our wildest dream, to the degree that it becomes 'magic'!"

"Science is always magic Dexter! To know is the greatest thing! What can I tell you? I command you to know! Can I make sense saying that?"

………………………………..

Halloway took a sip of coffee. He was watching Sheryl at the bar and his partner was watching him. Sheryl was watching the news on television. It was an update on the bombing of a maternity ward at a downtown hospital that day. The attack left over a dozen mothers and expectant mothers dead. As usual, they took the babies and fetuses. That was typical of the Faith Defenders. These Christian zealots usually took the fetuses as physical evidence for their 'commander' – whoever he was.

She shook her head distastefully and changed the channel. There was another report about another attack. This time the media released the identities of fifteen healthcare workers who collected bribes to give out information on pregnant women. Sheryl laughed ironically and dropped the remote on to the counter.

"This shit is wearing me down Charley Boy!" Halloway hissed disgustedly.

The youngster bit into his pepperoni pizza and drawled back at the big man, "Yeah, I know what you mean. I feel the hell too! She's a high-profile case. We must find her...that is, if she's still alive!"

Halloway paused and then replied tensely, "She's still alive. But...these days...there are too many toxic psychos out there! Those metal brained assholes will do any goddamned thing!" He took another sip of coffee and watched Charley with a one-eyed glare. Halloway seemed strangely nervous.

The youngster meditated on his pepperoni pizza, a sober look on his face. That thought was inevitable. Marilyn could be dead. Halloway seemed desperate for denial. Was the lion heart feeling the hot for a woman he pretended to dislike? That was far-fetched. Sheryl and Halloway were two of a kind.

Charley took a sip from his beer jug and leaned back comfortably in the lounge. "We have a tough motherfucker on our hands this time Halloway! The entire city is up our butt holes, expecting miracles! They want us to find her yesterday! It's been nine weeks! Why would you think she's not dead? How's she off the grid?"

"She's not dead! If she were, we'd still have to find her anyway! And we must find the culprit who took her too!"

Charley shook his head discretely, watching his partner. What a pretender! "You're assuming that she did not disappear on her own?"

"I don't believe she took herself off the grid Charley. How could she?"

"How would I know?"

"You wouldn't!" Halloway replied.

"You're right. I wouldn't. Like I wouldn't know why we can't get information on Faith Defenders people! Why are they off the grid? Some influential person works with them! If someone in high office isn't involved in Marilyn's abduction, then it must be Aduna Gyasi!"

Charley stuffed the last bit of pizza into his mouth. Then he emptied the beer jug and rested it down with emphasized finality. The tough cop frowned, "You don't wanna have a couple more beers with me? It's still early."

"Ten fifteen is passed my bedtime partner!"

"Pussy! Go home to Mama!"

"Sure I'm going Halloway! To my wife and two kids! You need a life! Who drinks coffee ten o'clock at night anyway?"

Halloway held his cup aloft to scrutinize it. He turned his blood-shot eyes to Sheryl at the counter. "*Me!*"

"Yeah – You! Think about it! You need to settle down!"

"Why?"

"Why not?"

"Because it makes no sense! I tried it before –
remember? Now I'm problem free!"

"If problem free means you drink coffee night and
day then you got problems Halloway!"

"That's because you assume that I don't sleep."

"You don't sleep! Red-eyed irritating pain in the
neck! Marry Sheryl! The two of you are peas in a pod!"

"We're alike! You're so dumb boy!"

Charley put out his palms apologetically, "OK! You
have a point. She's a female John Halloway. You still
need a life!"

"I have a life! My only problem is my job, your
nagging like a bitch...and right now, finding Marilyn
Trooper is a pain in the ass!"

"We'll find her! We must!" Charley affirmed,
punching the table with stubborn avidity. He rose to his
feet and added, "And you'll be the hero! You'll marry
her and sleep in your bed at nights...I mean, in her bed.
You'd love that, wouldn't you?"

"You're talking out of your ass again! Just because I
believe she's alive!"

Charley chuckled, knocked the table with his palm to
say adios and turned to go. John's tough face betrayed
disappointment, but only for the briefest instant. That

was most unusual for him. Something was on his mind but Charley could not stay.

"Go home to your wife kid. We got work tomorrow! Say hi to Elaine for me!"

The big man watched his partner leave. Charley waved to Sheryl. She waved back at him, a perplexed look on her face as she pointed her nose in Halloway's direction. Charley nodded in acknowledgment.

Halloway knew what they were telling each other. Sheryl argued with him about it before! It made no difference to him. He knew what he wanted, and he needed nothing but what he had! Besides, he always said he did not give a shit!

The oddest-looking pair sat across to his right. The woman, a six-foot tall African-American, slender built and dressed in full white. She was not that shapely, but she had enough curves to make her attractive. She wore very light makeup, so that one could appreciate the effect of a naturally beautiful face. He was Korean or Chinese, about five feet and a couple inches tall. He had an attaché case, which he held dearly as if preventing it from running away. He was an executive...maybe. The man was trying to convince his partner about something. They had been at it all night! Her incredulous grin certified that he was not making a good case for himself. Maybe he did not know how to tell the perfect lie! Halloway smiled and turned his attention to the entrance.

A gorgeous blond haired woman in black came in. He was watching her discretely when the couple stood up, distracting him. The tall woman wanted to leave, but her partner was trying to persuade her to stay. Sheryl saw them and went to collect. The man paid up – with cash! Then he hurried off to catch up with his unhappy date.

Halloway smiled. All the years they had tried to get rid of cash money, it never waned in popularity! Tangible cash in hand felt good! It would be some time before everything went to the chip.

He watched the strange woman move gracefully to the corner from which the couple had left. It was closer to the door, and she could look out into that snowy Boston night. She seemed like a blond woman, but even in the dim light, he wondered if she was Caucasian. Everything was everything these days that it almost seemed hypocritical to play the race game. You could end up hating yourself in the process! Aduna Gyasi was right when he talked about the phallus and vagina prevailing over petty prejudices. Curiosity always screwed the cat!

There was something contagious about her. Something he had not seen in any female for a very long time...until recently.

This woman had real curves on, and they were built firmly too! She was lean, but not too thin. He loved 'wire waist' and broad hips, stiff pointed nipples and tight 'cocked' butt. *'Here's some rich body to hold on to!'* he thought to himself. The super cop held his face down and smiled at his own smutty-mindedness. If Charley could see inside his head now, he would never have another day of peace!

The woman took the corner seat facing him. Her legs fell apart, enough to say 'hello' without seeming lewd or deliberate. He looked up again and it caught him off-guard. The first thing he saw was the forbidden view. Halloway's eyes were stuck unwittingly, but he managed to drag them away, hoping to spare the embarrassment. He raised his head to face level. Their eyes held and locked. That brazen woman was challenging him! The cop looked away. He was not the one in the mini skirt!

She smiled mischievously. Her brown eyes caught the amber chandelier light and took a sparkle. Halloway saw sweet champagne dreams, glaring in those crystals. He felt like he would suffocate.

No, she was not Caucasian. She was *mixed*...like himself! History recorded that the 1990's through 2000's were the days of the 'sexplorers'. Nothing could stop the orgy! By 2013, the new paradigm became a norm, and even the heavens declared it. This period proved frustrating for the KKK. They stepped up their campaign for new recruits but not many of them were pureblooded. It compelled a redefinition of what it meant to be Caucasian. That never happened, as it would mean conceding the fact. For curiosity, humanity transformed like a caterpillar, against all their resistance. Some saw beauty while others saw abomination. Most purebred claims proved to be hypocrisy. Two thousand and seventeen was unofficially the year of zebras. By that time, *everything was everything!*

With the cyber-baby concept all that happened in the past would threaten to re-occur. There would be new notions of blood-lineage, which would evolve into race, race wars and a competition for the right to rule. The god-creators, and even Aduna Gyasi himself, would likely choose those that represented their legacies to rule the others. It could even be worse if, as it were in the beginning, these god-creators created the first prototype from their own DNA. They would certainly love their children more! This kind of prejudice would defeat the purpose of evolution, creating stagnation and regression. That is, if it were not for the irrepressible phallus and vagina.

Evolution was to make Mankind advance to as strong as possible. No one mixed concrete with cement alone. The strongest man possessed the best of all humans.

He turned his attention to Sheryl. She was watching the news on television. That, he did not want to hear! They were talking about the missing reporter!

Halloway stole a glance at the mystery woman. She was watching too, which gave him the chance to view her unnoticed. She had a nicely sculpted body and that she knew! He harnessed his thoughts! It had been too long since he was with anyone!

Halloway felt her eyes bore into his forehead again; watching him watching her, and only God knew for how long! He turned away, feeling like a loser. The cop waved to Sheryl. She knew what he wanted.

In less than a minute, Sheryl came with his Budweiser. "Are you alright big boy?"

"Yes I am Sheryl," he replied hoarsely, "Couldn't be any better!"

Her eyes trailed discretely to the woman in the corner. "I hope you'll do something interesting tonight. You're living a pathetic life Halloway!"

She left before he could respond. He downed the beer in one go. Sheryl knew! That woman in the corner had his ticket, and he could not find the antidote! The last time a woman got to him, he reacted like an ass! Tonight he was open for it. Charley and Sheryl were right!

He looked up again. The woman was coming! He stiffened, feeling his heart leap to his throat! Halloway watched from the corner of his left eye. His right hand fumbled nervously for the beer jug on the table. It was empty!

He was not looking now, but he had her image fixed on his retina. She moved with an infectious rhythm, like a model on the catwalk. Her eyes were confident and her lips curved forward at the corners. She seemed intent on making mischief with him!

The cop braced himself preparing for her. He looked up in time to see her go by his table, heading to the bar! Halloway breathed easy and smiled at his own stupidity. He expected that gorgeous one to come at him! That would not happen! He had to go and talk to her! He was a turn off for so long, he forgot how to make conversation.

"It sucks, doesn't it?"

It was she! He tensed conspicuously but she pretended not to notice. The intruder put a jug of Budweiser in front of him. Then she sat down and took a sip from her own.

"W-What...d..?"

"Drink up!" she said, interrupting his attempt at a question.

He smiled awkwardly, and reached for the beer, "*Thanks*!"

"I was just saying...when you have no one to go home to – like me and you...it sucks – Right?"

He was about to ask how she knew that he had no one to go home to, but he changed his mind. "Yeah...kinda. I guess!"

"I know who you are!"

"Too many people do!"

"True! You're a *cop celebrity*! Shows how screwed up our heads can be. It's the toxic thing – Metal brained generations!"

Halloway frowned, "Well, that couldn't be a compliment! Hope it's not an insult!"

"Well it's neither. Just a common sense non-toxic observation."

Halloway chuckled. She caught his eyes and asked, "Are you getting closer to finding that reporter?"

"Marilyn Trooper? *Naw*! If you know anything, now is the time to declare it!"

"If I had a clue you'd be the first to know! Everybody loves her!"

"And I have orders to find the woman that everyone loves!"

"So...you don't think she's dead by now?"

"Naw..."

"Why?"

"Don't know. Just a gut feeling."

"A gut feeling?"

She eyed him suspiciously. "You got a crush on her, don't you?"

Halloway was about to speak. He paused to make sure about his answer. Her eyes pierced him inquisitively. Then he offered, "Maybe...Why?"

"You seem overly concerned; like it's personal to you. That's all."

"My job is personal to me."

"O...K. Your job is personal!"

She watched him muse over her sarcasm and then shrugged with no apology. This woman shot straight! That was something he missed every day. Maybe it was a generation gap. She was quite young. They fell silent for a while. He looked across at Sheryl. She gave him the thumbs up.

"You live around here Halloway?"

"I didn't hear you..."

"You don't live inside the bar, do you?"

"No I don't!"

"Good! So where do you live?"

"Sorry kid, I can't recall!"

She pouted stubbornly, feigning jealousy. "I know your problem! You're fretting over the reporter! I bet she can't fuck like me! Let's take me to your house and prove it!"

Halloway reached for his jug and took a sip of beer. She poked her face at him analytically, "You look shocked Mr. Halloway!"

"Are you crazy?" he blurted, confused. "I don't even know your name!"

"Well, do you want to fuck me, or do you want to fuck my name? Pick one!"

Halloway swallowed more beer and declared, "I like nameless people!"

She giggled infectiously, "I hope so!"

"Yeah...and you know what?"

"What?"

"It's coming back...like I can recall!"

"Recall what?"

"My address. I remember it now," he confirmed.

"Maybe you took your supplements today! It's good for your memory!"

Halloway chuckled.

"Do you have a Meter at home John Halloway? I don't want to catch the claps!"

"I know!" he said comprehendingly and then added; "One man I knew caught the claps and sent in an application to euthanize himself. Authorization took months to arrive, and he died the day the papers came through!"

The woman shuddered with revulsion. "He should have taken a bullet! Screw human authority! Retarded shits! They don't want you to live and they don't want you to fucking die! The only thing they want you to do is stay in limbo feeling pain! That's why we have all these metal-heads running around!"

....

Icilda rifled through her bag for a Meter. "Gimme you' wrist muscle man!"

Sam complied and she strapped it on him. He watched digits run across the face of the gadget. Finally, as he expected, it rested on negative for all counts. She took it off, and handed it over for him to return the favor. She was clean, and not even ovulating, which was a major plus! Samuel Powell lived for unprotected sex!

It was four weeks now since Jane left, and he could not get used to the single life. That night he went to the bar 'looking'. What he was looking for found him first! She was a fine Jamaican girl, who his herculean frame impressed.

He stood by the mirror, staring at his image, like a fan of his own. Strangely, he could not fully appraise himself from a reflection. He saw the muscles though! They twisted like steel cables around his six-foot body. He did not look much different from his college days when he was star quarterback. Those were his glory years! He won a basketball scholarship at his high school in Jamaica and ended up being a football star instead!

Icilda was short, even for a woman. He liked her neatly curved body. Even when she was not trying to impress, she naturally took on the right strut that turned him on. Her blouse did not cover her flat abdomen. Just looking at her belly button kept him horny the entire time. He was happy with his luck that night!

She stood by the side of the bed hustling to take off her clothes while making side-glances at him.

"Police boy!" she agitated with paradoxical affection, "Hurry and take off you' clothes!"

Sam hurried, throwing the pieces all over the room. Finally, he took off his boxers, which ended up on top of the dresser mirror.

For some strange reason, his own image distracted him again. Icilda was impatient. She went over, took him by the hand and said, "Come now man!"

Sam mused at her boldness. Icilda was a liberated human being! She went backward, pulling him with her. He pushed her gently and she fell onto the bed. Immediately, she fondled for it and inserted his organ inside of her before even kissing him. Foreplay became an after-thought!

She was passionate! He could not believe how firm she was! With everything he did to curtail it, he felt himself on the verge of ejaculation too soon. He stopped his breath, but that did not help. Her sensation overwhelmed Sam! Her muscles felt like a mouth inside of her sucking, 'blowing him off'! "*Oh...!*" he grunted unwittingly, and she half-giggled.

It was sacrilege to deflate prematurely with a Jamaican woman in the house! Sam braced himself. It did not help either. He had to let go!

"Ohhhh *shit!*"

Icilda giggled at his frustration. "Big man, you finish already? You a rabbit?" she teased, as if she did not know what she did to him.

"This isn't normal for me!" he said, hoping to convince her.

"You selfish!" she replied, but he saw that she was joking.

"Sorry babes! This isn't me!"

She rubbed his head sympathetically. "No worry! Not every man can handle me!"

That was a merciless poke! This had never happened in all his life! He would prove it the second time, and he was almost ready again!

Then two shots rang out in rapid succession! The first entered the room through the window. It passed over their bodies and crashed into the mirror! The other bullet hit someone outside! The horrified female victim screamed in pain. Icilda trembled at the wretched tone of her voice.

Instantly, the lamed stud disappeared and the super cop took over! Icilda's eyes bulged in their sockets as she kicked reflexively, to catch her lost breath. Her scream could not happen! Sam cupped his lion palm over her mouth. He looked into her eyes to remind her who he was. Icilda nodded affirmation to his unspoken words and relaxed. Sam removed his hand from her face and whispered softly, "Roll onto the floor!"

She obeyed him, moving with such agility that he heard no impact when her body hit the ground. He loved this girl! She was afraid, but rational, playing her part.

He stepped off the bed and over her, crouching toward the door. Samuel had a gun in his hand, and she wondered where it came from. He peeked through the window. Three Faith Defenders were attacking a pregnant woman and her husband!

"I'll call the Police!" Icilda whispered, but he did not hear her.

The big man eased silently through the door and out into the snow, where bullets were firing. He had no clothes on at all! Icilda remembered why she was there, and not even the fear of death could be more significant. She caught her breath, pulled her legs apart, and reached for the extractor.

Samuel Powell deposited a load! She had turned the instrument on, so it would make him over-excited and ejaculate quickly. Now she wished she did not do that! Outside of her duty to humanity, she could have allowed herself to enjoy him! He was very hard and handsome. Two things that were missing in her life!

Samuel hit the cold with naked butt and bare feet that sank into the four-inch deep snow. It was below 30 degrees, but he was too hot to realize it in the moment.

They had on typical white robes, with hoods to hide their faces. One of them approached the scene from the street and the other two were with the couple.

The terrified man was begging them to spare the life of his pregnant wife, whom they had already shot. She was bleeding from somewhere around the ribcage, and laboring to get up. Her husband tried to use his body as a human shield for her, but the second man clobbered him hard with the butt of his pistol. He fell face down into the snow. The original shooter pointed his weapon at the victim, to finish her off before gutting her. Samuel had to act! He could not reason with devils. From the corner of his right eye, he saw the man from the street react to his presence, raising his pistol to fire!

Sam took a dive, stiffening in unwanted fear as the shot whizzed by his head. Ice particles bit into his unprotected skin. Then he realized what he had done! He was freezing, and naked at that! The muscle man smiled ruefully. He fired, but not at his attacker! He hit the zealot who was about to finish the woman. The bullet tore into the villain's stomach. The man folded over and the gun slowly fell from his grasp. His assistant seemed too confused to react. Sam rolled, shifting his body to change position, bringing back his focus to the man who fired at him. The attacker fired again, but Sam's sudden shift took him off target for a second time.

In a flash, the super cop was sitting up to face him once more. He caught the Faith Defender off-guard. The Christian Soldier could not re-adjust his position in time! He screamed, anticipating sudden death. Samuel fired twice. The first one hit the target in the stomach, slowing him down. The other punched a neat hole between his eyes. Then Sam shifted his position yet again, wondering why the second man had not fired at him...but he was the coward of the lot! He had no confidence in surviving a shootout. Sam brought himself back to a sitting position. The man was running down the street, his white robe flowing behind him like Dracula's cape.

Samuel shook his head disgustedly. "They will run and get away...because you cannot shoot them in the back!"

He rose to his feet, positioned himself, aimed and fired once. A red-gash appeared, as the bullet tore off the back of the culprit's skull.

Sam's body felt numb! His feet were freezing! The cop charged for the house shouting for Icilda. Behind him, the victim's husband was pleading desperately for help.

"Icilda!" There was no sign of her. Sam went to check the bathroom. "Where is she?"

Then he remembered the injured woman outside and declared, "Oh fuck that!"

Sam smiled gratefully. The numbness was gone! He had only been out there for seconds! The cop hustled to put on warm clothes and boots. Then he went back to check on her. She was not bleeding badly, and he heard sirens coming in the distance. Icilda had called the cops after all! At least she had a heart!

"She'll be alright!" he said to the man in an upbeat tone. "Just hold some pressure here for her until the medics arrive. Alright?"

"Alright s-sir!" the young man replied, encouraged by Sam's confidence. "Thank you so much! They came to our house down the street! We were trying to escape!"

"Now you must relocate! Don't communicate with friends or family!"

Icilda stood at the back of the house. Her heart pounded so hard, she felt like collapsing. She did it! *"Secret agent girl!"* she sang in a whisper.

She stood out in the cold without much clothes on. Fear got the better of her and she would not risk going back into the house. When he went outside, she grabbed her jacket and went through the back door. Her coat only covered down to her waist. She left her skirt and panties inside! In the end, he was not the only one out butt naked! At least she had her boots on! Barefoot in the snow would be impossible for her!

She pulled the lid on the extractor and dropped it into her jacket pocket. That would give them the location. It made better sense to meet them down the road, because the area would soon be crawling with cops!

Icilda walked across the snowy back yard through the neighbors' property. She even put on the swagger, although she knew her fanny was showing! Who cared? She was on the most important mission! A light came on to expose her. A man watched from a side window. *Whatever*! Icilda gave that house, and all the other ones around, the middle finger...for those who were looking at her fanny! "What have y'all done lately assholes?"

She crossed the lawn to the street. A black BMW SUV pulled up beside her and the door flew open. She hopped in beside Aduna Gyasi. Doo-Bug sped away. The wizard smiled at her bemused, "You love excitement! You should be a cop, or a soldier."

"Secret agent," she offered correctively.

James Carter sat in the back seat, eying her with nervous anticipation, until she pulled out the extractor. "Here Mr. Gyasi! Samuel Powell – The super cop!"

"Great!" Gyasi replied approvingly.

Icilda seemed apologetic, but no one cared to ask why she was naked. Carter handed his coat to her. "Fine specimen – I can tell you!" the biologist confirmed.

Gyasi nodded and turned his attention to Icilda, "I see you had 'challenges'!"

Icilda looked away bashfully. "It's a worthy sacrifice. I met a really good man too!"

"I see child!"

"Hey! Did you hear the shots? 'Cause things got crazy over there! Sam took out a couple of Faith Defenders! Left my stupid clothes behind!"

....

She pounded against his body rodeo style, working him hard but hoping he would not ejaculate quickly. He did not and she had an orgasm! Where she came from, those were hard to come by! He watched her excite herself and it intensified his own high. She half-screamed before her climax, getting overly excited. He thought he would have to hold on to her...but she calmed down, looked into his eyes and smiled. He smiled back. She leaned forward and pecked his lips. "Mr. Halloway!" she mused in a breathless awestruck whisper, "It was a pleasure to fuck you! Now that you deserve to know, the name is *Hope*!" She reached her hand out and he accepted graciously.

"Nice to meet you Hope! The pleasure was all mine!"

Halloway turned her over and got on top. The woman half-screamed again, with ecstasy. She wrongfully assumed that he was done, when he was just beginning!

The big cop finally made it. She enjoyed feeling his body break submissively, deflating against her. Those moments always reminded her of a woman's power!

"*Gee!*" she said gleefully, "Congratulation Mr. Halloway! You got your stud pass!"

"I should be congratulating you!" he informed her between gulps for air.

She patted his shoulder, "Ease up off me! I have to go to the bathroom!"

He rolled off to the side and she went. What an experience it was! He was worried about how well he would do after so long! Things turned out all right!

In the bathroom, Hope reached between her legs for the extractor. She had turned off the sensitizer because she wanted to have fun with Halloway. Had she kept it on she would not have achieved that rare orgasm. She closed the top to alert the crew. The specimen was all ready to go.

Halloway saw her come out, in a rush, gathering her clothes. He frowned, "I thought you would stay with me tonight."

The big man sounded boyish in his disappointment. She giggled, deliberately messing with his head. "Why? I don't live here! We came to fuck and we fucked!"

He raised himself up and sat on the side of the bed. This was unbelievable! He felt like begging too! He could not imagine sleeping that night without some more of her. "I am asking you to stay though. Please."

She walked over to him, cupped his chin and kissed his lips. "I can't believe this! John Halloway is dying for more of me, after playing hard to get in the bar!"

She let go off his chin, pulled away and went through the door without another word. He watched her. Her triumphant behind, firm as ever, playing the victory rhythm.

Hope's high boots knocked wood as she stepped down from the porch to the step. She hurried through the gate and up the road to the Broad Way intersection. The black BMW pulled up beside her. The door flew open and she hopped inside. "So, it goes without saying! I got it!"

She sat down. Aduna Gyasi and Carter James said 'Great!' almost in one voice. Icilda giggled, "That's how great we are sister!"

"I know! We make the big boys happy!" She poked her head behind her as if looking for someone else inside the car. "That tough cop has a crush on Marilyn Trooper! You know that missing reporter?"

"*Marilyn Trooper*. Who doesn't know *her*?"

"She's been missing for so long. Most likely she's dead!"

"Maybe…"

"This guy isn't going to stop looking for her! He's obsessed I think!"

………………………..

In the days of Faith Defenders, the worst fear was getting pregnant! When Alexia saw her positive result, she knew her life would come down to a piece of bread! What would she do? She continued working at the office. Even after three months, she was strapping her stomach. It began to show before she realized it. Faith Defenders attacked her one Friday evening on her way from work!

She was exiting Logan's Tunnel on her way to Chelsea when she noticed a red Honda Accord behind her. She thought it looked 'nice', but the driver was tailgating her as if trying to push her faster. "Why are you in such a rush!" she muttered to herself.

Alexia pulled out of the left lane and cut her speed from eighty-five to between fifty and sixty. She drove for another ten minutes then took her exit on the right lane. The car was coming behind her! She frowned. "This guy drives like a lunatic! How fast was he going to cross lanes and take the exit behind me?"

She pulled off the main road into Nutty Joe's parking lot to get cashews. As she exited her vehicle, the red car pulled up next to her! Now she was suspicious! Alexia stood by her car door for a while, watching discretely to see who would come out of that vehicle. The door remained closed. She could not see through the tinted window. Alexia closed up and headed inside the shop.

She went straight to the aisle, grabbed the nuts quickly and got to the counter. "How much?"

"Awe Alexia! You know!"

The tall woman giggled, "I really don't remember! Give me change from a hundred!"

"*A hundred*?" The woman took the money smiling and rifled for change. "You buy it every other day my dear. You need to be more focused! It's twenty bucks!"

"I think it's because I usually buy more than one item," Alexia glanced behind her to the parking area. "So how's your day going Millie?"

"It's as good as a day gets in these times! With two suns outside, what more can one expect? How are you?"

Alexia looked behind her again. "I am fine."

"Good!"

"I believe someone's following me."

"*What?*"

"That red car in the parking lot. I believe they're following me."

"*Why?*"

"I don't know Millie. I'm in two minds…wondering if I should call the cops!"

"I would!" the chubby cashier offered, scanning the customer inquisitively. "By the way you're showing now, you should be concerned. About the Faith Defenders I mean!"

Alexia looked down at herself in disbelief. "I'm showing?"

"Yup!" Millie confirmed and then allowed the shocked woman to gather her bearings. "As a matter of fact, don't call the cops! You don't know who will come! Get into your car and speed through the unauthorized area at the back! I'm gonna call Melvin now. He'll let you through and then close the gate!"

"Y-You think that's wise?"

"I know that the Faith Defenders have people in the police force! That's all I'm saying! When you get home, pack your things and go to Timbuktu! They're on you now!"

"B-But I'm not even sure it's *them*!"

"It won't hurt to think worst case scenario! Or, do you wanna wait to know?"

"No I don't." The tall ginger haired woman nodded at the cashier. Millie took up the phone and made the call. Alexia backed away shook her head for 'thank you' and headed off. The red Honda Accord was still there!

As she entered her Pajero, the sun took aim at the windscreen across from her and she saw the white hood! Alexia took a deep breath, determined to stay calm. She slid into the driver's seat, closed the door, started up and eased out of the parking spot slowly. Again, she eyed the rear-view mirror. The car was not moving. She drove off as if to go for the main entrance. Now it was moving!

Alexia floored the accelerator and cut across suddenly to the back of the building. She saw the red car screech to a halt, as the driver prepared to reverse and navigate the turn on ice. By that time, she was out through the back gate and it was closing behind her!

She got home, packed what she needed and headed out for a remote corner in Falls River! The Faith Defenders would have all her information by then! She had to avoid family at all cost, for they would use her loved ones to track her!

The father of Alexia's unborn child owned the cabin on his grandmother's property in Falls River. He took her there once before he died. It was comfortable enough for her but that would only be as long as supplies lasted! What would she do then? She bailed out her bank accounts for cash money. Fortunately, she had saved really well for years!

When she got to the secluded property, she went to park at the back. Then she walked around, fumbling for the keys that her dying lover left her months before. She had happy memories there! Alexia turned the lock and pushed the door open. A little man with long white hair and beard sat on the floor! She screamed hysterically, and he watched her, amused. She stopped to catch her breath and was about to go again! "Calm down lady!" he advised with mocked agitation, "Do I look dangerous to you?"

"Some dangerous people don't look dangerous! What are you doing here anyway?"

"*Me*? I've lived here for months now! What're you doing here?"

"I'm here because it's my place!"

"After all these months, I clean the cobwebs, and you wanna just waltz in to lay claim? Where's the gratitude?"

"What gratitude? What I do with my property has nothing to do with you!"

"Sure it does – Obviously! This is where I live lady!"

She breathed in to prevent hyperventilating. "Who're you by the way?"

He reached his little eagle's claw hand forward, "The name's Ray lady! Ray as in sunshine!"

"Well Ray, nice to meet you, but I need my property now! Sorry!"

The white-bearded fellow rose to his feet, scanning her impolitely, "Sure you need your property! It's obvious why you're here. I bet you can do martial arts, and you're armed with fire-power. Those you must have to protect that protruding tummy!"

She paused, scrutinizing him suspiciously. "Are you one of those Chinese 'silver fox' people? You dress in that pajama thing and…your white eyebrows and hair."

"This is my Bombay clothes lady! I don't wear a 'pajama thing', and I'm not Chinese, but I do protection work!" He grabbed an apple from the bag she was holding and put a crunching bite into it. "You need me! That's enough to pay my keep! Besides, Jaggaths last wish was for me to protect you! He told you to move out here months ago…but you're too stubborn, dragging your feet!"

15

"I wish we'd do it the cave man way, like growing the shit in some stone age body part facility and then stitching it on! We could chop off the handicapped limbs and stick robotic parts on to their stumps! That would work boss! In 2013, they were still proud of that kind of fart! The face would still look disfigured, but that is comely for Weasel!"

Doo-Bug was wheeling around the lab, working with unmatched efficiency and complaining at the same time. Weasel watched him spitefully but with a hint of fear. Doo-Bug would never afford to make his life less than miserable!

"Doo-Bug!" Aduna Gyasi cautioned, seeing the worry on the wiry man's face. "Take it easy with him! That is your coworker and this is not 2013!"

Doo-Bug reached for a tube with a copper colored mercuric looking fluid. "Alright boss! We'll treat him like he's human! I'll just pretend that his ass deserves it!"

Weasel took a deep breath and asked, "What is that thing Mr. Gyasi sir?"

"This is the new and superior addition to your blood. It will bring back your 'embryonic' faculty for growth and re-growth. Humans tend to lose the best of what they are before they become adults. This milk is a reminder to your body!"

Doo-Bug poked the tube forward menacingly "Drink up Weasel!"

Weasel eyed the bottle nervously. "Drink it?"

Gyasi grabbed the tube from Doo-Bug's grasp. "Stop terrorizing the guy and setup the drip Doo-Bug!"

"My bad! Sometimes you just can't resist!"

Doo-Bug set up the drip. He plugged the needle into Weasel's arm, watching the wiry man bite his lips to contain his fear. Weasel tried hard to stay as macho as possible.

"OK Mr. Weasel! As unworthy as you are, you're going to look like somebody now!"

Weasel felt the exhilaration charge through his body. This was his reason to be there! Gyasi had faithfully promised him a life! He was a patient man but he could not help being over-anxious now. "When Mr. Gyasi?"

"What Weasel?"

"When will I be whole?"

"It will take half the time that it takes a tooth to grow Weasel."

"Oh...I see..." his voice trailed off because of his emotion. Aduna Gyasi smiled, feeling good for him. The god-creator delivered even greater news. "It is highly likely that it could go faster than that though!"

…………………………

His bare feet made light impressions in the eight-inch deep snow. Although he was a little man, it seemed incredible…but then, a man barefoot on a snowy day, on thick ice was bordering on impossible. He wore a white cotton outfit that would be nice for the tropics! The shirt was sleeveless and the pants were only ankle high.

Alexia often teased that he had on pajamas. Technically, he was supposed to freeze to death!

He went to gather firewood earlier that day. When he came back she was gone, and he knew she left on her own. She toyed with a dangerous idea the day before! It was about two hundred meters from her cabin to the little country road. That day, he saw her looking through the window to the distance. "No one ever comes to this place! A car passes like every two hours, and every time they are the same ones, most likely belonging to a handful of residents that live nearby."

"Jaggath built the cabin here because it's a remote place."

"That's what I mean! I'm sure it's safe to take a walk out. It would do me good!"

"You should hold that walk for a few months Alexia. Don't let anyone see you!"

"Why? People walk by every now and again. It's *normal*! No one will care!"

"Do you ever see pregnant women out there?"

"No…"

"That's my point! A pregnant woman walking alone anywhere is not normal!"

He was moving fast, gliding effortlessly across the ice. As he approached the snowy street, he frowned, wondering if he should go east or west. Then he saw them at the corner to his left! They were not official Faith Defenders, just ordinary rogues looking for a payday. Everyone knew how to contact the zealots to sell a fetus!

Ray was in time! He even saw her attackers before she did! She was walking back to the track that led up to the cabin. An old, dirty looking, pickup truck slowed from behind her and started to pullover. She noticed Ray and smiled mischievously, but the little man was looking beyond her at the danger. She turned around and saw them!

Two men stepped out from either side of the vehicle. They both wore bowling hats, and raw bearskins that they turned into winter coats. The driver was huge that to her he seemed like a bear himself. He came out with a shotgun in his hand, while the other fellow held an old bowie knife. They were hunters about to capitalize on the opportunity to catch a prey. Any prey that made good money was good to them!

Alexia stepped back nervously. Then she froze, paralyzed with fear! What they intended to do to her would not be nice! She remembered the protector and turned to catch Ray's eyes. He saw her soul, begging and apologizing without words. Alexia did not want to die!

The big man saw Ray and cackled. "Sunny! Look over there!"

The other fellow made a little chuckle too, "Well everybody wants gutting today!"

"I know! What you think we should do?"

"Give him what he wants and…" The little man stopped and tore his eyes wide in shock. "Jesus! Paul…This boy here has hardly any skins on…and he's barefoot at that! How the fuck..?"

Paul poked his head forward inquisitively. "I don't know why he's not freezing! Ain't he human?"

"Let's find out! You take that shotgun and plug a couple holes in his ass to test him! That'll let us know!"

The big man beamed at his partner as if he just discovered a genius. "Sure! You're smart today!" He took aim at Ray and Ray was not there anymore!

"Jesus!" Sunny exclaimed. He heard a slight sound, like static, but he could not pin a location to it.

"What duh fuck!" the bearded giant muttered in bewilderment, spinning the barrel around in search of his target. Even Alexia was affected by the impossible move. She turned around to search the bushes, but Ray was standing behind the shotgun fellow! His body took on a slight bluish glow, before getting back to normal!

The little man reached out and touched Paul's spine. Then he offered in a very kind oriental tone that sounded girlish, "Here I am mister."

"What duh fuck!" the big man muttered again, trying to pivot in a haste, but something stuck.

"You can't do it sir. You're dying slowly, and you're about to feel many pains."

The little knife handler faced Ray now, wondering if he should attack the magician. He decided! He lunged forth! As Sunny jabbed forward, Ray stepped back nonchalantly, shouting to the terrified woman, "Alexia, go home now!"

Alexia did not move. She stayed to watch the fight that ended in a few seconds! Sunny took a swing at

Ray's face. The protector stepped back to avoid it. While the attackers hand went around completing the arc, Ray stepped forward again and touched his shoulder blade. Then he stood up and walked away as if nothing had happened. He looked up and saw Alexia. "Woman! Are you still there! I told you to go home!"

Alexia was oblivious to what he was saying. She pointed to the two men frozen in place and asked, "What happened to them?"

"Nerve damage."

"Nerve damage?"

"Yes. They're feeling a lot of pain, but they cannot scream. They cannot move. They are dying."

She saw tears well up in the giant's eyes and then run over. Goose pimples started on her skin. "Oh my goodness! Get me away from this sight! I'm glad I'm alive Ray! Thanks for my life!"

As she finished those words, Ray screamed at her "Get down!" but she was not one to listen to instructions. An arrow struck her in the back! Ray turned to see the third man who had been inside the truck. He was about to fire off another! This time, at the little warrior! "You people don't learn!" Ray said in a cold bitter way. All the kindness was gone from his voice and it made his attacker fear him more!

The timid fellow dropped the weapon and said "OK! Let's forget it!"

"Forget *what*? You just shot Alexia! Don't expect mercy!"

It was a remote area in Shrewsbury Ma., on a cold winter morning! Gyasi told her about the property three weeks before, and she waited all that time to get there. It was one of his most secret possessions. A large parcel of land that was not registered in his name. As soon as they stepped out of the car, she felt like enjoying to breathe. The multiple sounds of exuberant life poured forth upon the drum of her ears all at once; yet, the chaos they made was nothing but music. Even her desperate angry heart softened a notch to appreciate the beauty of nature. She wanted to say wow moment words...but not to her ungodly father's benefit!

"Beautiful! Is it not my dear?" the old man uttered in his proper and boring tone. He wanted to appease her, but she refused to accept the olive branch. She was still deciding on what to make of him. The process would take time and effort. For now, she was not feeling overly tolerant.

Marilyn had on nice strong leather boots, and thick jeans, which he went out and bought for her. She wore them only because she could not dare postpone the trip! Even though they looked very good on her, she could have bought her own damn clothes!

He pointed to a huge boulder at the top of a slope. "There's the border of our land."

Gyasi started up to it and she hustled behind, surprised at how agile he was. The old man moved like a twenty year old! She was laboring behind him!

As they reached the borderline, he stopped at the landmark to let her catch her breath. It was a mighty stone, split in two, with each half on either side like a gigantic open door. Light snow dusted down on her face.

It was slowly covering the tips of threes and high places. She looked around her, heard the birds singing and felt the natural exhilaration fueled from enthusiastic forest life.

The scientist treaded forward again for her to follow. Then he stopped, looking down cautiously. She came up to him and saw what it was. A long winding precarious track, sloping down and incredibly steep! Her heart leaped to her throat, but she managed to quiet it down enough to pretend...that all he could see in her was fearlessness. She halted abruptly, fighting to catch her balance. When he held his hand out to her she latched on to it in spite of herself.

"Be careful, it is a rugged terrain!" Gyasi advised uneasily.

Marilyn pouted spitefully. "You've given me the runaround for three long weeks Dr. Gyasi! Don't tell me about the terrain! I know how to walk and we're not turning back this time!"

"No! We are not turning back!" he assured her. "This is it. Just prepare yourself!"

"I am prepared, '*Mr. Daddy*'!"

Aduna knew that he was paying for past sins against the daughter he loved so dearly. She was the only one that he had to alienate. It would take time and effort on his part to begin to heal the breach. He had done her a grave injustice but it was a sacrifice to humanity. He did not only deprive her. He deprived himself too. One day, as long as she was his daughter, she would become one with the project. She had to! She was the only one he had to depend on who had his blood inside of her.

Marilyn slipped and he braced himself as her legs came away from underneath her. The little man stood like a rock planted in the ground and she was amazed. How could he be so strong for a little fellow?

"Thank you," she croaked. He helped her back to her feet. Aduna turned his nose to the trail again. She followed; this time more cautious than ever and holding on to him for dear life.

They arrived at the bottom and she breathed a sigh. It was a little forest! Mighty pine trees pointed to the heavens and she even noticed a few maples as well. She enjoyed the sound of the birds and her heart warmed watching squirrels play chasing games. She had never seen nature like this before! Gyasi pushed forward again. She followed.

"Beautiful! Is it not?" he uttered as if he had not said that before.

Yes...It is very beautiful," she offered half-heartedly.

They were going through the woods, and it seemed like there was no end to it. Suddenly, they were in the clear on many acres of wide-open plain. The wind blew a chill down her spine. She could see the house! Her mother would be there! Without even knowing why, she broke away from his grip and charged off, in desperate need of what would be in store! Gyasi called out to her twice, but she was not going to take heed.

She got to within a couple hundred feet and then froze, paralyzed. Now she needed him to help her cope! He held out his hand to her but she did not take it. Marilyn fell into his embrace, like a very long-familiar daughter, taking him by surprise.

"Father, I should have asked you before. Why did she have to be 'dead' too?"

He felt the emotions surge through his body that it was difficult for him to speak. She called him 'Father' this time, and it was not to mock him! He forced his mind back into focus, took in a deep breath and said, "Because of Angel."

"I don't understand!" she said, squeezing into his body as if looking for shelter.

"Don't worry my child. You will," he offered in a choked up tone. That served to reawaken her resentment.

She released him, and it was like that positive emotional moment did not take place. Aduna Gyasi was a god-creator. He knew what it was. She was a smart one too, but emotional pain was no good thing to endure. He allowed her to pull away and 'estrange' herself.

They stood side by side, watching the house, waiting for some specific time, when some internal spirit would move their feet forward. It was a tense moment...a decisive one. Gyasi was not just introducing a long-lost daughter to her mother. He was introducing a long lost mother to her daughter too!

He turned to steal a look at her, questioning himself. Was she ready? She felt his eyes and looked around, catching him at it. He was depending on that intuitive spirit.

"*What*?" she questioned flatly.

"Nothing my child...Are you ready?"

"I said yes to you a thousand times!"

He focused on the house and stepped forward. She followed him. Marilyn loved that little house! She always lived in cities, not by any choice of hers. Imagine if she had grown up there in the middle of nowhere! Youthful days running around these expanses and watching squirrels play would be nice. A rowdy brother would present good insight into men! She was clueless as an alien when it came to the opposite sex! Her diet seemed to worsen her incapability to socialize too. People saw vegetarians as party crashers!

Marilyn heard a whizzing sound! She hesitated, eying the god-creator who went on doggedly, oblivious to it. Something moved at the side of the house! She strained her eyes to see it. Maybe she was imagining things! Nothing could move so incredibly fast!

Gyasi eyed her, "It's OK dear! Angel has his pranks!"

"*Who?*" she asked, needing no answer. Gyasi continued forward. Someone touched her shoulder, startling her! It was a pale finger, colorless as a jellyfish! She saw it, but then it disappeared into thin air. "Father! I-I...saw…"

She was looking desperately across her shoulder, but Gyasi's steady hand pushed her forward. "I said it's OK. Do not be afraid!"

"B-But I saw..." she was talking and turning at the same time. When she focused ahead, it was on one knee in front of her, sniffing! She wanted to scream, but held it back. Her trust was in Aduna Gyasi the god-creator. This was his handiwork!

"What did you do here Father?" she asked in a reprimanding tone.

Aduna Gyasi pulled his beard with one hand and scratched his head with the other, trying to figure an explanation. It was futile! "Meet Angel...your other part."

"My *'other part'*!" she returned sarcastically, watching the subject watching her. "I bet you mean to say my 'twin'! You have a lot of explaining to do!"

The little man shrugged childishly. She turned back to Angel. What an awesome sight! Angel reached for her hand and she eyed Gyasi, who nodded. The brother put her finger to his little pointed nose that had no visible nostrils. "What is he doing?"

"He knows who you are by your scent and whatever is wrong with you. Maybe you caught a virus."

"I'm not *sick*!"

"Let us argue about that later!"

Angel let her go and rose to his feet. He was over seven feet tall! The only thing that seemed real about him was his eyes. They were pearl-black and shiny, like Gyasi's own...and like hers too. His body was as transparent jelly. It felt 'liquid', warm and cool – You could not tell! He had no ears, no mouth and not much of a face. Sometimes Angel's skin would glow. His veins were electrical circuits, firing perpetually, especially in his head.

"Wish I could do that!"

The strange being did not answer her. She did not know what to think or do. Right then, standing before this alien of a human, she could only feel its love and her heart reciprocated. "Brother...I see we're not exactly 'identical' twins though!"

Angel did not speak! His intelligence was superior. Not even the great Aduna Gyasi could communicate with him! Angel's energy felt like 'love'. Marilyn wanted to embrace him but he saw no meaning in what she was proposing to do. She needed to connect with him badly! They would have to communicate! She would have to learn!

Angel turned suddenly and in a flash, he was a hundred meters away, heading for the forest. "Balls of Inky!" Marilyn exclaimed. "That's not possible!"

"It happened! He will bring your medicine."

"Medicine? Why – I'm not sick!"

"It will do you good to take whatever Angel gives! He is not a meaningless human."

"Meaning is not normal in humans these days! Without a mouth, does my meaningful brother eat?"

"No. Food would bring pollution to him."

"*Ahah*! So...as we are twins, why am I not fasting forever? I get the pollution - Right?"

Gyasi was trying to decide whether to answer her but she asked, "Why doesn't he starve to death anyway? I mean...wouldn't it be normal?"

He replied soberly, "You have much to learn my daughter. There is substance all around you...in space. How much does a tree in a pot eat in a year? Does it have a mouth? Food is no monopoly on substance! Come! Let us go see your mother."

"Well...silly of me to even ask!" Marilyn agreed sarcastically. Then she inquired tensely, "Is my mother normal? I wish you don't shoot back a question in response!"

"Why would you conceive such a question child?"

"Great! I take that for a 'yes'!"

"Anne Eve! My baby!"

A little woman came charging forth from the house. Marilyn froze in amazement! Apart from a few gray hairs at the sides of her head, Dianne Gyasi was near identical to the young reporter! The only difference was that the youngster had Gyasi's charcoal eyes! So why did her twin look like a stranger from another galaxy then? Life as a daughter of Aduna Gyasi was perpetually flabbergasting!

Marilyn had a terrible flu that night! She thought she would suffocate and die in her sleep. It felt like metal shavings were stuck in her throat. She could barely swallow, and she could hardly breathe. It was the misery of all miseries!

Marilyn propped up pillows to elevate herself. No way was she going to sleep in that condition. Maybe she needed to call someone! She resolved to go to her mother but it was easier thought than done. She was too weak.

Marilyn managed to put one foot over the side of the bed. Then she decided to catch her breath before getting up. She dosed off into a miserable half-sleep.

Someone appeared in the room...in her dreams. The room transformed into a hospital ward. A short black woman walked the aisle, looking at her patients. She was dressed like a nineteenth century slave. The woman wore African head-wrap and her spirit felt intensely angelic. Marilyn was in despair. She needed this divine being! At that point, the woman passed by her and winked. At the blink of that angel's eye, Marilyn's pains and runny nose vanished altogether! She was shocked out of her sleep. When she returned to this realm, she feared that the painful flu would come back! It was gone! She had experienced a miracle! Now she wanted to return and at least say thank you but everything had disappeared, including the door that had opened for her to outer-space.

In the morning, she was anxious to tell Dianne about her experience. As she opened her mouth to speak, the woman cut her off, "You went to bed sniffling last night! I can see that Angel brought your medicine!"

Marilyn's breath stopped for a wide-eyed moment of thrill. She clutched at her heart in shock. "Oh my goodness Mother! How could he do that?" she blurted, awestruck.

"Angel will marvel you my dear!"

"Marvelous is a futile word today ma'am!"

16

"Adversary!"

Gyasi was shocked out of his sleep. That voice sounded like cold thunder and it was hard to think he was dreaming. He rose up on his elbow and checked. Dianne was sleeping peacefully. He was sure he heard a voice. It was strong and mighty, like the sound of Niagara Falls.

Maybe he was dreaming then. He rested his head back down on the pillow.

"Look up satanic being!"

Gyasi went back up on his elbows. This time he saw the entity! It was human in appearance, but its face seemed horribly distorted. One side of its face was paralyzed, its eyes were ready to fall from their sockets and its lip-less mouth was twisted sideways.

The god-creator took a glance at Dianne again. She was still sleeping.

"We will not disturb her devil!" the being assured grudgingly. Then he added in a conspicuously disappointed tone, although with crackling laughter. *"Unfortunately, you are not afraid!"*

Aduna Gyasi replied, "No food for you! That hideous appearance does not work I suppose! You Watchers have no jurisdiction over the enlightened!"

"We understand that!"

"Why then are you here?"

The being lifted its ash-colored wings involuntarily. It pointed a wiry finger to the window. *"We must speak with you! Come with me!"*

"Why should I come?"

"So we can reason in a different dimension, not to disturb your spouse."

Aduna frowned suspiciously. Then he looked at Dianne. She was still fast asleep. He looked through the window and saw the little oval craft. "Why do you try to deceive me? You are in breach! Have respect for the Law! We are sleeping in our realm! There is no way she could be disturbed! I need not go anywhere with you!"

The being put forth a hand in truce, *"Alright devil! Fair enough! I respect the inherent Law in you! You must reciprocate!"*

"Where have I infringed on you?"

"With your ungodly project to enlighten our flocks and to usurp our world!"

"These souls are not yours! What other calamities do you have in store for them! Will you incite more wizards in a predilection for evil? Will Tutu Babe Coop not suffice?"

"So let them speak for themselves!"

"Fine! That principle goes both ways!"

"I know! That is why you sin against us! They have chosen anarchy of their own volition! It is no right of yours to choose otherwise for them. On this earth, those that are enlightened are free! We never dared molest

you! Those that are in darkness belong to us! How dare you decide in your arrogance to interfere with our design? Like our kind, you have become a god. Why don't you go and create your own world? Leave our own alone! Forget about the souls we claim!"

"You are right in the principle of free will." Gyasi conceded cleverly, "In that regard, let me put forth a solution."

"You shall not, demonic being! We will tell you what we require!"

"I will only agree to that which pleases me – within the Law!"

"Our demands respect the Law! Imbecile, we have served the Eternal from before you hatched! Henceforth, you will not seek those who deliberately refuse you! Ninety nine percent of the populations of the world hate you Aduna Gyasi! Of their own volition, they chose to see you as 'the beast of Revelation' and the 'antichrist'! You have no authority to force your philosophies upon them! They chose our prescribed notions of norm to cultivate as cultures and religions. They are not choosing your own! That gives us a legal right to take you up with the Law! Under the Law, we order you: Halt all public speeches and efforts to challenge our constituted notions of reality."

Gyasi chuckled. "I must agree with all reasonable demands affected by Law. I shall not deliberately teach those who refuse me. I shall speak in public, according to my legal right to express an opinion. I shall speak my philosophies, for it pleases me to utter them, and there is no breech in this! You must amend your ways, for it is a breech to enforce mediocrity."

"We never enforce mediocrity!" the being declared in defiance. *"We prescribe it and they allow it! We became their gods, by their own choice. They prostrate before us, by their own volition! They ask us for every definition instead of thinking with their minds, because of their own mediocrity! Our service to them is not free! We give them the definitions that please us, because they agree to deny themselves! Where in this have we made a breach demonic accuser? You must give up the Cyber-Baby Project! For we know what you intend to do is to usurp us via super-human faculties! We will not let you bastardize the prevailing human prototype!"*

Gyasi countered, "I afford you the power of the Law. It holds me guiltless to create my project among those who want it as much as I do, and those who, hearing and understanding, will be convinced."

The visitor replied with venomous zeal, *"We will fight you – within the Law – until you accept good reason to desist! If you do not cooperate with us, we will convince the humans to kill your bastard children before they hatch! You will reap the fruits of your labor! None of your blood will go up against us! Not one will die at our hands! Humans are reprobates, the scourges of the universe that you find reason to care for! They will serve you up to hell with the utmost joy! They will repay you hate for your love!"*

Gyasi chuckled again. "You will follow the rules, and so will I. I will not interfere with you, and neither should you with me. I shall carry on the Cyber-Baby Project, for it is my right. It pleases me!"

The angelic being hissed disgustedly. Its face turned to fiery rage. *"Then please yourself in the hell that will come from your ungodly Cyber-Baby Project! Obnoxious Satan! We will raise spirit against spirit! Your will in the bodies of Mankind against ours! Your legacies and credibility against those of the gods of Earth. You will see what attains when you kick against the prick!"*

"That is a fair challenge adversary! You stiff-necked demons and Satan of divinity will see the gates of your living hell break asunder!"

"Then this night is final! We have agreed to intensify the covert war, within the Law that stipulates human rights! You kidnapped Marilyn Trooper, but we have other news carriers!"

The being was gone! Gyasi fell flat on his back, breathing heavily and looking up at the ceiling. What ungodly beings these Watchers were! They controlled even the foods that the humans ate. They tried to turn these flocks into homosexuals to boycott the 'seeding' of Mankind with humankind. Therefore, they gave them synthetic foods with transgendered hormones. They attempted to make most humans infertile by feeding them destructive sweet tasting additives. These additives were to cause addiction and prevent fertility. Instead, the humans were breeding like rabbits anyway!

Now, in 2025, they were in a state of desperation, because the untamed phallus and vagina would soon be at it again! Unstoppable was the equation of God the Ethereal!

They knew that Dianne was alive, but they could not divulge the information to Tutu Babe. They did not know that Marilyn Trooper was Anne Eve! As it now stood, Tutu Babe would be his biggest concern! Tutu Babe murdered Dawn and Charm but he had inside help! Even if Babe knew that Ray was alive, it would not matter.

Dianne stirred beside him, and then turned dreamily. She threw her arm across his chest. "Aduna? You sleeping?"

"No."

"Good! I am still excited! Marilyn is even more beautiful in real life!"

"Television lied! She looks like you!"

Dianne wanted to talk about one thing when he wanted to think about the other! Aduna would ignore the threat from the Watchers and concentrate on fighting puny humans! That he should have done years ago! His daughters would have survived!

He did reap some degree of success. He would advance the human prototype. Anne was born to his legacy, the next project master. She was due for the great sacrifice that Lisa and Dianne accepted. The god-man was guiltless before the Law! For he did to others what he would have done to him. About Anne, maybe he failed in her faculties. He managed to keep her alive, which was supposed to be the hard part!

He created them as the first cyber babies – the two in one, which he had hidden from the world. Eve was the active prototype. She, as the Eve, would bear the first child. Angel was the mightier half, who passed for an insignificant side order. He found it necessary to split the embryo, expanding into a completely new experiment. The unknown came to being. It turned out that he could not communicate, but his superhuman abilities were a preview into what the Arpago would be. After years of research, Gyasi discovered that Angel was too much of a divine entity to comprehend lower forms as Mankind. He communicated telepathically, but no one could hear him. Every attempt Gyasi had made to discover any frequency range of communication in him failed.

If Angel was a true super mind, what about Eve? The super X chromosome was to manifest as a higher intellectual faculty in flesh. She was to run the operation, and become even greater than great God-Creator Gyasi. At twenty-eight, she was stuck in illusory realities! Did he fail on her?

The adversities that came with his early experiments were a serious setback to him. Now he was collecting what Carter James regarded as 'superior specimen'. They expected him to create the all-powerful hybrid from the best of these prototypes. He created what he needed decades ago! Creating what he did not need was important for protecting the project.

They recruited women prostitutes to collect specimen from people that Carter James found worthy. These men agreed to have intercourse with the women. They left their deposits without a care to the charge of these whores. Those deposits belonged to the authority of the women who collected them, not to the authority of the depositors.

Carter James was the best biologist among common people. He worked assiduously to select the best of Aduna Gyasi's sperms for the god-man to program. It gave him something to do and to believe!

The specimen that they collected now would create multiple prototypes in their volunteers. Carter would select the best of the best and neutralize the rest. These prototypes could only live for one year anyway. They were potent in their DNA, but as complete human entities, were nothing but docile blobs, just lying around until they died. Carter expected Gyasi to extract the chosen super blob for use in their final volunteer, the Eve.

Any woman could carry a cyber-baby. Only the specially sculpted womb of Eve could carry the seed of the cyber-baby. Anne Eve was prepared from conception! He checked her womb the morning after he abducted her. When Nurse Hudson did the internal scope, it caused her to dream about a sexual matter. Aduna Gyasi said he would use the seed of the man in her dream for her. It was a statement made to deceive. Anne was a sacred tabernacle among humans!

If she would not agree to partake in the mission, it would be death to the entire lab. Nothing was going to halt the process!

He remembered the best conversation he ever had with her. She indulged him that day. He asked her, "Tell me something about you that I would not know."

She giggled. He could tell that it was embarrassing for her. When he coaxed, she admitted, "I'm a bit of a freak!"

"Why do you say so?"

"I love to talk..."

"You don't talk much Marilyn!"

"I don't mean like *that*! I love to talk to spirits, in nature, like in trees! I consume the life force of the sun. In the winter it keeps me warm on cold nights!"

"How do you do these things? Give me one example!"

She looked down at her toes. "If I'm working hard all day, and my strength is drained, I find a powerful tree, and I put my palm against it. I greet it with love and befriend it, and I let the hollow in my palm rest against its mighty trunk. I talk aloud that it feels the vibrations of my will-to-be. I pray with words, asking it to share its energy with me and I believe it gives me strength. I never get tired afterward!"

"Feeding on electromagnetic force! Are you insane or awesome! Let us select a tree now and find out!"

She refused to go. That morning, before they left for Shrewsbury, he went alone and found a tree. It worked incredibly well! The energy went up through the hollow of his palm, right above the wrist. He felt incredible power even then. He remembered catching her up when she fell. His strength was from that energy! Marilyn had something in her! She needed to connect with Angel and his marvels. As soon as he sat down at the steering wheel she spoke questioned, "Is that why people don't like me?"

"What do you?"

"I mean that I'm a freak! You made me into a freak!"

"You are not a freak. Everybody likes you!"

"Name one, apart from you!"

Gyasi stopped talking and leaned his head sideways to think. She eyed him suspiciously and leaned her head sideways to reckon. Marilyn did not expect him to answer her question. He spoke again, "Your relationship with people is not affected by anything in your blood Anne Eve. It is the same for you as for anyone else!"

"Why are they not indulgent towards me? Is it because my parents abandoned me?"

"Growing without your parents affected you. Getting old without your daughter is equally distressing."

" *'Psyches'*! Right?"

Aduna stopped to think again. Then he nodded comprehensively and exhaled. "Anne Eve! People are uncomfortable around you!"

"Why?"

"For the same reason a naked person is uncomfortable around one who is dressed. They feel inadequate. To commoners, you seem incapable of a fault. You are insensitive, not knowing you intimidate them."

"I intimidate people? How? Could I intimidate a big strong cop like John Halloway?"

"Especially ones like him!"

"How and why Aduna Gyasi?"

"Because you seem perfect in your ways, you are better at what you do and you are more beautiful than anyone else!"

Marilyn looked surprised. She knew she was beautiful but not relatively perfect as he implied! It was a problem, but she felt a bit of pride in it! For the first time ever, she examined herself. He had a point! Marilyn swallowed her high air and then inquired humbly, "What can I do about me sir?"

"Make them feel important too! Common people are afraid of their own mediocrity, so they avoid the higher circles. Be sensitive!" He paused and leaned his head to think. Then he added, "This world confuses our kind. Your learned behaviors are incompatible. You are sufficient in yourself. Time will prove it! A man who touches the daughter of a god to profane her sanctuary must surely die! A woman who bears fruit for the son of a god shall be raised above the children of men! She will feed at the gods' table, for she was mother of a god, who suckled him!"

"Wow! Tone down!" she agitated shaking her head confused. "That was way too much, and I don't know what tangent we're on! What do you mean I'm sufficient? And who are children of gods and children of men?"

....

"This is where they raised the Iron Soldier back in the days!"

"I know!" Charley countered, "Although I was not around to fight against it. It gives me a creepy feeling just being here!"

"A creepy feeling? I don't get it boy! What gives you a creepy feeling?"

Charley raised his palm at the bigger cop in futility, "Alright Halloway! You don't understand! Forget it!"

Halloway frowned. "OK."

Charley was pacing around the large conference room, thinking. "Why are we even in here! There's nothing to find!"

"Don't know Charley! Maybe it was a prank call. We searched everywhere! I don't hear any voices! Do you?"

Aduna Gyasi and the lab abandoned the property after the state passed a stop order on the project. Pimps and drug dealers soon broke the locks. Now they used it for illegal operations. Investors were eyeing the building for business purposes. They could not contact Aduna Gyasi to make him offers. The state had to decide on what to do about that piece of real estate. It had become a security hazard!

Over the past weeks, a particular gang took over the turf and kept everyone else out. That morning they received a call from HQ to go and check out 'voices' screaming for help that were coming from the building. They received numerous 911 calls about it. One caller swore that the voice sounded like Marilyn Trooper's!

"Well, I guess we'll be leaving now! We checked! Marilyn's not in here!"

Charley put forth his hand suddenly like a traffic cop, "Shh!"

The big man's palm instinctively reached for his weapon and locked around its handle. Then he froze and listened. The wind from outside coming into the building made eerie noises. Sometimes it rattled loose wood or scraps of metal. He was just about satisfied that everything was in order when he heard something like a twig snapping under foot! That sound John Halloway would never get wrong!

In a flash, he pulled his firearm and turned to the direction of the sound. Charley was watching him puzzled. Finally, he pulled his own weapon and turned to face the opposite direction. Halloway kept his focus ahead and moved to where the sound came from.

Charley followed closely, looking the other way. As Halloway got closer, someone jumped from behind the wall and fired!

"Down Charley!" the big man screamed hitting the floor. Charley lay on his belly and swiveled around to see the threat. He heard Halloway fire twice. "I got him!"

The cops rose to their feet cautiously, scanning the building for the slightest move. Halloway walked boldly across to the still figure. It was a gang member, but he had on a white FD hood. "Well what do we have? This gang is affiliated to Faith Defenders it seems like! There must be something here!"

There was another man against the other wall and Charley saw him. He had no choice but to fire at the cops. As he fired, another man appeared from somewhere and charged out of the building! Charley fired at the shooter and missed. His eyes widened in panic when he realized that the tug was about to squeeze off another in their direction. He steadied his aim, knowing he took too long! Halloway spun in a split second and fired without aiming. Charley watched the thug fold over and fall down on his gun.

"Jesus Charley! Call for back up! These creatures pop up from everywhere!"

Charley made the call while Halloway checked the injured gunmen. They were dead, but the cops could not take comfort until backup arrived.

Suddenly, Charley exclaimed excitedly, "I know what Halloway! Why are there floorboards around the podium? Let's rip 'em up!"

Immediately, they heard the sounds. It sounded nothing like Marilyn Trooper! They were screams from at least a dozen tormented souls, but none of them had a US accent!

The backup came in five minutes and they began to work. They found a huge basement area underneath the wooden structure. Inside were at least two dozen wide-eyed black girls no more than sixteen years old or less!

17

After the cyber particle transfusion, Weasel did not want anyone to see him the 'old' way again. Gyasi kept him in isolation until the process was complete. Four weeks later, it was time for him to return to civilization! He was staying on the East Lab range, to the far eastern side, in one of the abandoned buildings. Weasel enjoyed it too! He could lie around, doing nothing all day, except watching the mirror to see how far he had advanced and then watching the Discovery Channel.

Aduna Gyasi went to pick up the lanky fellow that day. Although he was the wizard, the results astounded him! Carter was watching microscopic particles on big screen when he came in with Weasel. Weasel greeted the biologist, sounding as if he was fresh from a vacation. "Hello Dr. Carter! Long time no see sir!"

Carter lifted his head instinctively, "How're you doing mate?"

He turned back to face the screen, aiming to lose himself in concentration again. Then it occurred to him! He looked up again to scrutinize the wiry man. The ex-handicap chuckled to see the jaw dropping effect he was having on Carter.

"*Weasel*?"

"Yes sir!"

"What in bloody hell? *Weasel*?"

"It's me sir!"

Carter could not believe his eyes! Weasel even seemed taller! The biologist turned to his partner. "Aduna Gyasi! By the balls of Inky, tell me what it is mate!"

"His body healed, with a little help from alien intelligence."

"All life forms in the universe heal or advance by alien influences my friend. Just tell me what it is!"

"I took his DNA samples, schooled his genes and sent them as agents back into his body space to remind him how to morph, as he would do at the embryonic stage."

"*What*? Did you grow body parts..?"

"You are not *listening*. He grew his parts back, just like a child grows its teeth."

Carter shook his head in amazement. "Gyasi, I've been waiting to see manifestations like this one! You've always outdone me! My life work in biogenesis is redundant, but you still hold on to secrets! I grow replacement parts in a lab. You reprogram genes to remind the body how to grow parts! Do you know what this could do for us if you decide to sell to Tutu Babe?"

"We already had this conversation Carter. What will happen to the masses if a few become gods?"

Carter saw that his argument was opening old wounds. "You're right my friend!"

"I am right Carter!"

Carter frowned. "If we can alter humans with boosted genes, why is the cyber-baby project relevant? Genes transfusion could turn man into Arpago! You said that humans are more formidable than iron soldiers, as they can evolve *'to chew iron if necessary'*."

Aduna shook his head, disappointed. Power and greed were prohibitive to conscientious thinking! *"Impossible Carter! You do not have a clue about Arpago's faculties. I can create these gene particles, and they can be the next best thing to immortality. But, if you give physical might to spiritual decadence the result will be as the Iron Soldier!"*

"Arpago's power is not only physical. He is mighty in mind and spirit. Spirit faculties depend on the earth's energy field along with the cosmic. Their limit depends on the conditions and degrees of intensity that prevails. To Create Arpago, we build a vacuum in the womb of his parent. There we simulate conditions that are not available to the present age. Arpago is a future mind from a future light realm. It exists many thousands of years from now. At that time, the earth vibrates at a higher frequency and so does the body and the blood that runs through the vein. So does the mind and the fires of electrodes within the brain. At that degree of vibration, fire is ignited in the foot prints of man!"

"And man can be on fire or charged as an electric eel without burning?" Carter asked for the umpteenth time in the project.

"I can prove this, but it would only waste precious time!"

"You're never wrong Aduna! It's beyond me to figure 'how', so I keep asking. I'm looking at Weasel and it befuddles me!"

Gyasi nodded understandingly. Carter was thinking hard. He waited for the biologist to speak. "How do you make cyber genes that the body won't reject?"

"There is no synthetic stuff and nothing to reject. Does a man's body reject education? We are born robots operating on intelligence programmed in our genes! DNA is a program, and genes are programs. They are human particles, which I reprogrammed for specific functions in humans. In Weasel's case, I took his DNA sample, extracted the particles and reprogrammed them. Can you understand how the earth programs us? It turns us on, off, up and down by universal remotes? It gives animals seasonal behaviors."

Carter stuck up his finger like a student in class, fielding a question.

"Yes my friend."

"We often talk about higher frequency realms without considering the implications. If we vibrated at a higher frequency in our present bodies, we would die from high blood pressure among other things. I mean, while we perplex ourselves trying to send our minds into overdrive..."

"You have a point Carter. Everything goes faster in a higher realm! When you are single, you live in a little apartment. When you take a woman that space becomes cramped. You live there until you make it bigger or move to a better place. You live in the new place until the problem repeats, for now you have children. It is similar with our spirit beings. We live in this home while we improve on it, against perpetual resistances that are cause to good effect."

"This phenomenon you have seen and will continue to see in life. You talk about high blood pressure. I tell

you this. Every now and again, the wizards will return to you and say 'I think normal is higher than we first thought'. When you go back in time, you will see how far it moved. This is perpetual motion. We must do our deeds faster to secure longer lives. Longer lives lead to greater wisdom."

Carter smiled ruefully. "You lost me again! The number of years you live determines whether you lived long. Why do you talk about it in terms of us doing things faster?"

"Time is mere illusion created around events. Time does not create events. One second and one hour are relative notions. It depends on the events that we use to measure the interval. For example, how many thoughts you think between the time of sunrise and sunset, or to what degree you will be rejuvenated in a second's break. This is what brings value to any notion of time. It is not how long the sun takes to rise and set. That is backward thinking!"

"When we do more between events, our lives between them are longer. If you walk a mile to the farm every day, that length is a value against your workday. If you begin to drive, your day becomes longer by that degree. You live longer in that day. Notice a ray of light coming from its source. At, and closer to source is the bottleneck. It is a smaller space, but smaller beings have the same relative notion of its size as do larger beings at the wider end of the funnel. The event of sunrise and sunset would seem longer at the wider end, yet their days would be just as long at the bottleneck, because they vibrate at a higher frequency. One day for those at the wider funnel will be many for the being in the bottleneck. Imagine now, if the beings at the wider funnel begins to vibrate as fast as those at the bottleneck. Their lives would be vastly enriched, because

the value of their days would be multiplied. This will be our Arpago!"

"O…K. I do think I have it now. You used the word 'immortality' earlier. Do you think we will achieve it?"

"What good is it? I said 'the closest we could get to immortality'! The concept of immortality is for soulless beings!"

"Why do you mean my friend?"

"A soulless being has no faculty to conceive better! Mortals incarnate to advance, not to immortalize. We study the caterpillar but we do not see a lesson in it! Will it stagnate in its generic state, or will it transfigure? Humans are eternal soul beings. That which is immortal is trapped in one dimension. That which is eternal transcends time and paradigms. Your home is not in lanes and tunnels but lanes and tunnels are to take you home. Remember your past lives, conception and birth, and come face to face with your eternity. When you know that you are eternal, you have no interest in the monotony of what would be immortal."

"I understand." Carter went back to his project. Gyasi watched him knowingly. Then the god-creator spoke. "Tutu Babe will live long on earth as a vampire oppressing the innocent! His torment will be worse than that of the innocent he preys upon! He will not live long enough to justify the sacrifices he makes either. Not even fifty years above the average mortality rate!"

Carter turned to face the god-man, looking uncertain of himself, *"What?"*

"As we were talking about immortality, I remembered Tutu Babe. He has scientists working to make him immortal. What sense does this 'Faith Defenders' organization make? How many true Christians are involved? Where do the children and fetuses go? I am sure the churches do not use them!"

"So...you believe that Tutu Babe is responsible for all that?"

"You mean sacrificing human lives under the guise of Christian zealousness? Well, what do you think?"

....

Pete loved listening to the kids speak. He never missed 'Timeout for Kids' unless he did not have a choice. Today they were interviewing a five year old boy.

"What do you want to be when you grow up Sidney?"

"A male mother ma'am!"

"*What*?"

"I want to be a male mother, like my dad! He's pregnant and he said being knocked up is *cool*!"

"He's just joking about being pregnant!" the woman said tensely, opting for damage control.

"Nope, he's knocked up!"

"O...K...but, what kind of job do you want to do?" she replied, trying to steer the boy away from exposing the subject.

"I'll just take care of the children, like my dad. Chuck works at the harbor and my dad stays home."

"Alright son! Let us not say any more about this. You have a right to be whatever you wish to be…and you have a right to change your mind too…I mean, if you choose to…"

Pete chuckled! Now he only had to contact Jennifer for their information. A fetus was a fetus, even if it was inside of some gay person!

At 9:30 PM, they went to the house in Mattapan. The kid was at his mother's place. His father was home alone as his 'other half' was one to stay out late at the bar. That made it easy for Pete.

They were on the first floor, and Pete beckoned Frank to come over to where he was. "How do we get this rat inside?"

Frank shrugged and went to the side of the house. After a few minutes, he called out to his crony. "Come over here! Bingo!"

Pete hurried over to Frank. A window at the side of the building was open! "Great!"

"You sure he has a cat Pete?"

"Jennifer asked the kid and he said yes!"

"Alright then, put the rat in!"

Pete raised the box with the rat inside. It was bigger than any cat he had ever seen, but that was what rats were coming to! These giants would attack their feline foes on sight. The rat went in!

After a minute, the racket started! A man was screaming girlishly! Pete looked through the window.

The person had on women's nightclothes and he was jumping about hysterically. They went to knock on the door. "Hello in there! Is something wrong?"

"*Eeh*! Help! A giant rat!"

"Open up and let us see!"

In a flash, the door flew open and the man appeared with curlers in his hair. "It's attacking my cat! It's killing my cat!"

Frank walked in with the box and whistled. The rat stopped what it was doing and went back to captivity. The girlish man watched him. Then he asked, "Is it yours?"

"Yes it is! It got away!"

"Be careful next time! Alright?"

"Alright!" Pete chipped in offering his hand. "I'm Pedro! Nice to meet you!" He was watching the man's protruding stomach as he spoke.

"My name is Floyd Cupcakes. Nice to meet you too! As you already noticed, I'm extending my family!"

"That reminds me! I came here about *that* in question!"

Frank grabbed the 'stylish' fellow from behind, grinning. "Let me go brute!" the pregnant male shrieked, kicking desperately to free himself. "My man will kill you if you harm me!" His threat held no conviction, and Frank cackled like a hyena.

"Sorry *Cupcakes*! I must gut you regardless," Pete informed him with mocked apology. "Take no offence at all! Just a job for me you see!"

"Just a job for you? My man will find you! He'll gut you back and shit on you!"

Pete was oblivious to the threat. He went to help Frank secure the victim and tape his mouth shut. When they finished 'straight-jacketing' him, Pete took the container from his carrying bag. He reached for his all-purpose dagger and pierced Cupcake's stomach.

18

For two weeks, Marilyn stayed in Shrewsbury with Angel and her mother. Gyasi went back to work at the East Lab hideout. He besieged her to use that time to decide on becoming a part of the project. That meant giving up on the life of Marilyn Trooper and picking up where she left off as Anne Eve long before she was old enough to know.

They were anxious weeks for Aduna and for Dianne who sacrificed so much. What if she refused? She grew up on prescribed conventional norms. They had no influence on her life. They hoped what was inside of her would make her accept in the end. The secret was safe when Anne Eve Gyasi did not know who she was. Ironically, Marilyn Trooper was ever eager to uproot and expose them! Now that Anne Eve was Marilyn Trooper, who would prevail?

From foster care to foster home, they made sure she was all right financially. That was all they could do without risking the infant's life. To use project recruits as foster parents for her would be imprudent. Project recruits were the cause of her dilemma from day one! Her foster parents were not supposed to know a thing. Neither could she. That was the way to protect a secret from Tutu Babe. They watched her from a distance. She had a mystery godfather who never came by himself. It was good for her foster parents, because they benefitted well too.

Dianne bonded well with her long-lost daughter. As much as Gyasi felt *Marilyn's* wrath for his sins against *Anne*, Dianne went free of charge! She knew that it only meant the opposite of what it seemed. Anne was most upset with her dad because she needed him most. With all Anne Eve's indulgence to her, Dianne could only gain brief moments of her time anyway. The reason for that was Angel. She spent hours a day sitting with him and even forgetting to eat. She seemed addicted to his presence.

On the second Friday, Anne rose at 5:30 AM. She went straight to her sibling. Dianne prepared breakfast, but her daughter did not come down even at 9:00 AM.

At 9:15 Dianne went to Angel's room. The door was open, but they were sitting together on the floor, both appearing to be in a trance! She left them and went back to her business, wondering if she should tell Aduna.

Even at 12:45 PM, Marilyn did not come down for breakfast. Dianne went to the room again. This time she stood at the door and called out to her. "Anne!"

The youngster did not seem to hear her, so she raised her voice just a bit. "Anne!"

Angel refocused. His thoughts came back to the realm. That released the young woman. Anne Eve was shocked out of the depths. She spun around blinking at Dianne, a confused look on her face. "*Huh*?"

"I did not mean to startle you. You haven't eaten at all since yesterday!"

"Don't worry mother. I have much to eat that's better than food!"

"*What*?"

"*Energy*! The room is full! Don't you feel it?"

"Not as I ought to..." Dianne surmised and took a left turn to make a hasty retreat. Something weird was happening between those two! She should say, between

Anne Eve and Anne Eve. That was a truer statement. It was confusing but everything Gyasi did was confusing. This was the mystery of all mysteries. They were no twin! They were one soul that depended on no other soul. If they were sibling, the coming result would be futile!

When Gyasi came to pick her up that day, Dianne pointed him to Angel's room. The door was ajar but they were sitting on the floor, doing nothing. Gyasi decided not to disturb them. Three hours later, after many checks, they were sitting the same way. He made up his mind and said, "Anne, it is time to go!"

Angel released her again. Anne saw Aduna. "I didn't know you were here!" she declared, sounding disappointed and rising with emphasized reluctance.

"I told you I would come by mid-day. It is 3:30 now."

"Is it? I thought it was around eight in the morning!"

"That is off by a margin Anne!" Gyasi mused.

"I was out of this world!"

"Why do you say that?"

She giggled defenseless. "I'm your daughter! I communicate, telepathically, with Angel! He showed me marvelous things!"

"I thought he was indecipherable!"

"To your gadgets – Not to my mind!"

"I am baffled. We tried all possible ranges to communicate and got nowhere!"

"You should have tried the *impossible* ones! What kind of scientist are you anyway? I am learning more than any human on earth, apart from Angel himself!" Marilyn touched her head. "Deep down in here there's a place of silence. There, words are obsolete and nothing exists. Your mind is generic consciousness determining to 'become'. You start from the first inherent notion. A dot becomes a circle. From one circle, you conceive many. Where two circles overlap, a straight line occurs. It is the beginning of conjecture. You imagine, creating the holograph of being. From straight lines come angles and from angles come diverse shapes. It is art. Art creates all; for it appreciates instead of rationalize. Art is creation – There's nothing else. Science is essential to cultivate dimensions. Art is transcendent. You have forgotten that one thing. How then could you hear Angel? I communicate with him through art creations. Even your art becomes a limitation of science. Our art has no dimensional logic. It is not a ballet...it is the reggae dance. I conceive his designs and he conceives mine. I told him the things you told me, that I could not understand. He explained them to me. He explained your shortcomings too!"

"My shortcomings? Like what?"

She ignored his question. "One thing I know! I'm not going to lose Angel again!"

"You will not if you become one with the project."

"Mr. Gyasi...Father...I have no choice. You created me this way. It'll not end until somebody put together the things you broke!"

Gyasi was relieved that she was 'in'. He was careful not to show his excitement too much. Her wisecrack drew an unwitting chuckle from him though. He was Aduna Gyasi! She did not really believe he broke things! He 'created' things!

Anne sighed and then asked, "What're you planning to do with me? I'm your daughter and guinea pig at the same time! That's disgusting!"

"You are my daughter! You are no guinea pig!"

"I am too! Let's face it!"

"You are not! You must separate the issues! If you were, then all creation would be! Can you not comprehend it?"

She put her hand out like a traffic cop. "I can't comprehend crap! I accept what I am, but I still don't know what that is! What's done is done! Tell me what's next!"

Gyasi breathed in deeply. "We have the specimen. You will become Eve for many generations of humanity!"

"I'm not sure what I will be Father! It burdens my heart! Will I be Eve for perfect generations, or will I be mother of all abominations?"

"You are a creature of Perpetual Motion. I am its stalwart. You are my child!"

She turned to her brother. "And Angel? What about him?"

"Angel is my son/daughter."

"*Son/daughter?*"

"Never mind! He is my son!"

"I know he's your son. But what is his purpose in our project?"

"The dimensional mind of man ever questions the supreme god self. His purpose is to know all and see all. With what he is inside of you, you will be the one."

"Will be? That means I'm not! By the balls of Inky, what confusion!"

"Eve, you and Angel are like radios. Your soul is one frequency from the same channel. How many radios are there?"

"Two."

"And how many channels?"

"One; but I still don't get what you're saying!"

"It will take time."

"You made him into a useless experiment!"

"We tested cyber genes that will become Arpago inherent."

"Which confirms he's an experiment!"

"No Anne! He is the embodiment of your subconscious – The glory of you."

Marilyn bowed her head sadly, *"The glory of me?* What can he do on Earth? This is not his paradigm! He cannot be here!"

"No human really believes that he belongs here. We all have a subconscious mind – a higher self. None of us had ever seen what it has to deal with before *now*."

"You love to contradict!"

Aduna Gyasi was about to say something but changed his mind. Naturally, she was getting to know Angel better than anyone did. She thought he did a wicked deed by creating him 'different'. She was wrong! He did not make Angel different. He made *her* different! Her difference manifested as Angel. It was difficult for a conventional mind to understand. The concept of one in two bodies went far beyond normal reckoning. She understood that if it were her leg then it would be attached to her. She would not realize a higher definition of attachment. She never saw the space between the very things that touched, or the distance between atom shell and core. Every time he tried to explain, she would simply revert to her first confusion. He would leave it to time.

Gyasi was stroking his beard thinking when she stated, "He's coming with us!"

"Why? He stays with Dianne!"

"I'm not going anywhere without him!"

"He will be with you. You will see!"

As they headed through the difficult track to the car, Anne remembered a puzzle. It troubled her mind since the moment she met Angel. She stopped as if it was impossible to mention while walking. Gyasi sensed that she was not coming and halted in front of her. Then he turned to face her. "What is it my dear?"

"Is Angel a light being?"

"This question draws paradoxical answers that tend to confuse. I will not say that he is a light being but, for the same reason, I will not say a caterpillar is a butterfly. A caterpillar is equipped to endure a process that ends in it becoming a butterfly. When Angel becomes a whole body in itself, it will be a true light being."

"You do not see him as a whole body and yet you can't see the wrong you did to him! Moreover, you just referred to him as 'it'! By the way, where do I fit in all of this? He and I are twins!"

"What is more important to you than your soul? You refer to the same as 'it'! Angel is the higher you. Whatever he is, you are and what he becomes you become. He cannot advance without advancing you."

"Another puzzle. Right?"

"Right!"

She smiled spitefully and then said with a sarcastic undertone, "Since I'm what he is, let me ask you! Am I a light being?"

"You can take the above answer and put it in that slot!"

"*Really?*" she screeched, thinking that he was being sarcastic too. "Why can't you give me clear answers?"

"I give you clear answers when I have them. I leave you to question yourself when the answer lies in you."

"I questioned me all day! Now that I realize that I don't know the answer, I ask you!"

Gyasi turned to go without answering her. She hopped up beside him determinedly. When she spoke, her eagerness to learn touched him. "What makes a light-being?"

Gyasi froze. "A light-being consist of only what is real. You saw it, spoke about it, but by the time you get back to reason you forget!"

"Are you saying that people are not real?"

"We are real because the one thing that is real indulges in conjecture and authorizes existence. We are the images of light and its energies. Below the surface of the earth there is light, a huge ball of flame. It causes all matter to stay in the order. It is the Sun of Earth; our Mother Sun. Above us, there is light. It causes all matter to conceive. It is Our Father Sun. In the galaxies are other forms of light casting different shadows. All these lights, greater ones and lesser ones, create frequencies, greater ones and lesser ones. All these frequencies create images, greater ones and lesser ones. We are the images, coming from darkness through light onto reckoning. The only thing that is real is the light and the intensity of light. A light being exist in the knowledge of light. It indulges in 'imagination', but it understands the reality, is never far from knowing, and its predominant truth is 'I Am All that I Am'. It sees itself from within and without. Therefore, it cannot be infected by the ulterior."

Marilyn was astounded. "I only grasped a wee bit, but it leaves me breathless! It sounds like we have a legacy to god-hood! Are you telling me that we can possibly attain to that degree of omnipotence?"

"I am telling you to be aware of what you are!"

19

September 9, 2025; 11:00 AM...

A strange amphibian vessel rose up from the sea to the shore of the Ivory Coast. The villagers watched flabbergasted. Their prudent ones moved to quick action. It did not look good! They had to warn the authorities!

While they debated with friends, making a quiz over it, the machine sent forth an arm at the top. This transformed into a funnel-shaped object about three feet in diameter. It was an amplifier of sorts, pointed directly at the village! The craft stopped in place, and they watched it with open fear.

Inside the vessel, Samuel Uche eyed the commander frowning. "Are we in America?"

In a mocked show of friendship, Captain Dale Saunders clobbered the scientist's back with his big leathery left palm. Uche grimaced, but not for pain. The obligatory indulgence to this bastard was repugnant to him!

Saunders grinned with suppressed malice. "We could be *anywhere* but in America! I thought it was obvious! America was our departure point! Now isn't that some real common sense for you? Or don't you see the logic?"

"I must say, your logic gets me every time!" Uche shot back disgustedly, "How much nonsense can a single person make?"

Dale Saunders was a feisty and arrogant ex-navy captain who figured he was wiser than everyone with darker skin! He was furry as a gorilla and the rest of the boys called him *'Hairy Dale'*.

"Alright *'Who-She'* or whatever your name is *'Doctor'*! We left America hours ago! Now you're in the Ivory Coast – On land! It's your big day! You get to witness your machine at work!"

"Are we going to the forest?"

"To the forest? For what? Them niggers don't live in the forest these days!"

"Whatever slurs you use, it only confirms that people live around here."

"That's the idea! The gadget's meant to kill *people*! We must test it on *people*!"

"No way! Tutu Babe said we were going to some animal reserve in North Carolina!"

Hairy Dale laughed raucously, "You thought he didn't know that it wasn't the same with animals – Right? Let's start with the niggers out there!"

Uche was about to prolong the debate, but then he changed his mind to concede, "Alright! I'm already here! Murder the entire village and let me watch!"

"Now you sound like a partner…*Partner*!"

The scientist spun on his heels. Hairy Dale asked suspiciously, "Where are you going?"

"To fucking wait for you to pull the button already so I can go home!"

Hairy Dale chuckled, "You're right! Let's go!"

Uche hurried ahead of him. As they got to the compartment, the scientist entered the code in a hurry. The control chamber opened. Hairy Dale waved him aside distrustfully. "Don't touch anything! I'll do it myself!"

Uche pulled back his fingers and smiled, "Alright! I won't touch the controls!"

Outside, the village people gathered to watch. In the distance, what seemed like an army vehicle made its way to the coastline. The amplifier went all the way to the right. Then it returned in a semicircular motion, sweeping the village with a high-pitched unbearable sound that tickled the leaves of the coconut trees. The oncoming army lorry stopped and the soldiers were in frenzy, desperately clutching their ears. Some fell off the truck on to the street.

On the beach, it stunned the people. They held the sides of their heads in a hopeless stupor! They swaggered back and forth in frantic attempts to muffle the deafening sound. The machine continued to sweep the coastline. Uche watched people go down like pin balls, for as far as his eyes could see…

Hairy Dale and his crew were screaming, but with awe-filled delight.

"Satisfied now?" Uche asked matter-of-factually.

"Oh sure! You're the man 'Who-She'! You're the man!"

He punched the switch and the amplifier withdrew, disappearing into the roof of the craft. The machine took to reverse, swiveled to one-eighty degrees and went back down into the sea.

Then…slowly, the would-be-dead people on the beach began to move. Some writhed in pain, some were throwing up…and a few even began to rise lethargically…

Uche watched the captain break the good news to Tutu Babe. He listened nonchalantly to their insults. Neither did they care what he heard. He was just a slave to them. Otherwise, he was insignificant. Race mixing drove the KKK to irrelevance. Tutu Babe was ready to redefine what it was to be Caucasian! Everything in life was relative!

"This machine is the most awesome piece of equipment I ever laid my hands on! You should have seen it! Niggers were dying and no property destroyed! The entire village is dead! All fucking dead!"

"Uche told me that it has a range of at least a mile. Does that seem accurate to you?"

"I believe so! They had army people coming from around half that far. Those niggers suffered as much as the ones on the beach!"

"Splendid! We must reward Uche!" Tutu Babe chuckled in a cheeky way and added, "Black presidents and scientists are always useful to us. This is one example why!"

"What do you mean boss?"

"I mean, 'Black' Obama murdered Gaddafi, and nigger Uche murdered that village in Africa!"

"That's why a black Jesus never comes! Four hundred million niggers crucify him every time!"

....

Carter was off for the day. It was a good opportunity for Gyasi to sneak below the basement and get something done in secret. He told no human about this underground lab. Doo-Bug was the only assistant he needed down there. Protecting the project would not be easy, and he had to make tough and smart decisions. The hidden compartment beneath the floorboards of the basement led down into a tunnel. That tunnel went all the way across the acres to his property, where Dianne stayed over a mile away. No one would ever find it...but if anyone did, he would have a surprise for that person!

He watched the robot pull the incubator out and then start to calibrate it. It tossed the template across the wooden table. Gyasi went to toy with it for no reason at all. Doo-Bug came with robotic efficiency, took it from the god-creator and started putting on the meshing. With a scan of his eyes, he double-checked the dimensions to make sure that it perfectly matched the little man's size and shape.

Doo-Bug finished and went to install it into the incubator. It took up a finger thin plastic tube and held it in its right hand.

"Here's a little trivia for you! DNA Casting or cloning? Which is better?" the robot quizzed, distracting Aduna.

The little man agitated, "Doo-Bug? That is a silly question! How can you be asking what you should teach? You cannot compare these things. One is just a remote body. The other is the complete replica."

Gyasi was unsuspecting. The robot wheeled his way and raised its left hand. A sharp and deadly looking blade replaced its index finger simultaneously. Doo-Bug grabbed the scientist's left hand with its right, and with one swift motion, cut of Gyasi's pinky finger at the nearest joint.

"Oh fuuuckkk!" the wiry man screamed hysterically, more from the shock than from any pain.

"It *is* fuck to you sir! I can't believe you said that word!" Doo-Bug enjoyed the matter. It was still holding the scientist's hand. In one split second, it plugged the plastic tube over the affected stump of a finger. "You're lucky that fingers do re-grow!" the mischievous robot chirped, blinking with mocked malice.

Gyasi closed his eyes tightly. It would stop bleeding as soon as the tube was properly fixed. He waited for the imaginary pain to recede. Finally, the treatment started working and he managed to convince himself that he did not have to brace for agony. He regained his composure and his authority as well. "You will be lucky," he hissed begrudgingly, "if I do not scrap your ass by tomorrow!"

"*Sissy-pie!*"

"My daughter would so love you now! *Asshole!*"

"I know! Now you must explain a missing joint of pinky finger to her. Make up a story, because Doo-Bug never lies!"

"Doo-Bug never lies? Since when?"

Doo-Bug wheeled its way in front of Gyasi and asked matter-of-factly, "So you're insinuating that Doo-Bug is a liar?"

"I would love to emphasize it!"

Doo-Bug slanted its head and blinked its eyes in mocked disappointment. Then it asked, "Sir, are we going to do another cast? Because I'd cut off the other finger now!"

....

Anne Eve was inside her room watching the 'Timeout for Kids' when they interrupted the program. Something incredible happened on the Ivory Coast!

"A strange vessel came out of the sea and turned an amplifier on a small village! It emitted a high-pitched sound so intense that it almost drove the people insane. It affected the entire village, hospitalizing thousands! Thankfully, no one died."

"What is this? Where's my father?"

Anne sprang to her feet and went to look for Gyasi. She pulled the door open and almost bumped into him. He put his hand forward to hold her up and she saw that a part of his pinky finger was missing. "*Yikes*! What's up with that Daddy?" she quizzed painfully.

"My finger? Do not worry child. It is a life-saver and it will grow right back!"

Anne shrugged, willing herself to take courage. "Did you see the news? Some crazy thing happened on the Ivory Coast!"

"What was crazy about it?"

"An amphibian vessel came out of the water and emitted a sound wave that disoriented the people…the entire village of around eight thousand!"

"I see." Gyasi replied nonchalantly. She frowned, wondering if he did not realize what she had said. Aduna Gyasi was thinking. He patted her shoulder, turned around and walked off, "Come with me!"

"Where are we going?"

"To show you a jewel I made!"

"What jewel?"

"When you see it you will know! Wait here!"

She waited while he went to his room and returned with the thing in his coat pocket. Anne was anxious to see what it was, but she kept calm about it. Gyasi would do what he wanted when he wanted. They got to the general lab. No one else was inside.

The god-creator busied himself, and even seemed to have forgotten about her. She watched him setup a table against the western wall. He then went to a cupboard next to Carter's office and took out a dozen crystal glasses. Anne scratched her head but managed to contain herself. The god-man placed all the glasses in a line, allowing no more than two inches apart.

Then Gyasi came over to Anne. At least he had not forgotten her! He went to his pocket and pulled out a beautiful red stone that looked like a ruby. The jewel was about the size of both her thumbnails joined together. "Look!" he said offering the stone with pride. "Is it not a beauty?"

She admired the thing. He had made it into a pendant. "It's beautiful Father! I think you said you made this thing…but I figure you're joking – Right?"

"No my child! I made it long ago for you to wear!"

"I'd love to wear it, because it's gorgeous! I'm thinking that there's a reason to wear it apart from 'beauty'."

"It will protect you!"

"From what?"

"*Here*!" He plucked the jewel from her grasp and went back to the table. Gyasi placed it into one of the glasses. Then he went back fumbling inside the cupboard. He returned to her, carrying a strange-looking gun. It would seem like an ordinary weapon, if the nozzle were not a funnel expanding to six inches at the top. The god-man handed the gun to her. "Hold the trigger down and point it at the table."

Anne obliged. In three seconds, all the glasses exploded, except the one that contained the jewel!

"See? Go get your jewel now. It is to protect you!"

"Great testaments of Inky!" Anne mused in utter disbelief, handing the gun to Gyasi as if she were afraid of it. She went to the table to pluck her jewel from the glass. "Do you have a chain I can wear it on?"

Gyasi reached into his jacket. "Yes! I did buy one long ago as well…"

Anne suddenly turned to him wide-eyed, "That's what happened on the Ivory Coast!"

"Yes it is."

"It's not life threatening, is it?"

"You saw what happened to the glasses!"

"It makes people explode?"

"They 'burn' from within. But objects could explode too."

"L-Like w-what kind of objects?"

"Like your crystal glasses…and maybe your telephone!"

"My telephone?"

She blinked at him stunned, expecting a comforting answer. "Wear the jewel child," he said flatly. "Don't worry about crystal glasses or telephones that explode."

"B-But t-they did not k-kill anyone on the Ivory Coast though!"

Gyasi leaned his head. "That I realize…and it is strange. If Samuel Uche wanted them to die they would be dead!"

"Who is Samuel Uche?"

"A good person who foolishly got caught in the clutches of evil."

"So he was testing his weapon and making thousands sick? That's no good person!"

"I believe he was testing Tutu Babe's weapon that he made. That weapon was made to kill. If it only made them ill, it had to be Uche's fault! Somewhere in there, you will find 'goodness'."

"A merciful error deliberately made!"

20

In September 2025, they made seven insertions in the East Lab Shrewsbury hideout. Seven women got the honor to etch their names in the history of a world to come. Doo-Bug finished off with the last one, and wheeled himself out of the room without a word. Nurse Hudson took up the queue and gave Weasel the sign to tidy up.

James Carter watched from the big office. Aduna Gyasi stood behind him, beaming at their handiwork. "I am sure that any one of those prototypes will be just fine but, with seven to choose from, we will only take the finest!"

"I expect the finest to be from that cop fellow!"

Gyasi did not answer. He was watching Weasel intently, but for no obvious reason.

"Gyasi," Carter said on a sudden thought, "how is *she* dealing with all this?"

"*She* who?"

"Your daughter partner! Who else?"

Gyasi stroke his beard meditatively, turning his head sideways for thought. Then he said assuredly, "She likes to give me a hard time, but I think she is very much into it!"

"She's the only one who can stop us now."

Gyasi replied flatly, "You are wrong Carter. She is the best body available! The only one who can stop us is yours truly!"

Carter's mind was ticking conspicuously. He realized and stole an anxious glance at Gyasi. The god-creator seemed oblivious to him. Carter frowned, "You mean the project can go on without Anne?"

"Yes. The difference is that a substitute would cause me extra hours of hard work!"

"Well that's a revelation!"

"Why? I thought you would know! We will 'extract' the DNA prototypes after they are six months old. In two years, we will do the insertion for Eve. Nature will activate the programmed DNA inside of the prototype. From this, we will take genetic material to program the seed we choose for Eve. In this case, we could use that of Halloway's, which we will cultivate to the genetic extract from Samuel Powell's prototype."

"Magnificent!" Carter Mused. "When one hears it on the media, it sounds synthetic! When one sees your methods, it only seems natural…like something God himself would do. It's a perfect insight into how to create like a true god!"

Aduna Gyasi smiled, watching Doo-Bug wheel back into the room and flash a thumbs up at the camera.

"Your absolute finest creation!" Carter appraised.

"Do not be too sure about that," Aduna declared. His meaning was lost to Carter, who passed it off for a joke. All was going well to him!

"Tell me something my friend. How's that kid?"

"Who?"

"That cyber-baby kid that turned out useless – Anne's twin."

"He just hangs around."

"I see. Can he walk now?"

"A little...with a lot of help," Aduna lied. "It is putting a strain on Dianne!"

"She's wasting her life! Imagine the pain that thing is feeling too! Terminate it!"

"I cannot terminate our child!" Gyasi shot back flatly. Carter realized his insensitivity. "Gosh! You're right man! It does no harm to keep him around. It sucks that Dianne doesn't want to see her friends anymore! We were a team man!"

"We were a team that had to survive on trust. She has withdrawn from the world but that is fine. I watch her back and she watches mine."

"We all watch each other's back! That's why the project continues!"

"I guess so!"

....

September 11, 2025…

Samuel Uche was in trouble! He knew it well in advance! How he got involved with evil people was no mystery. Aduna Gyasi paid him well enough but he chose greed! Tutu Babes proposal seemed too good to be true. It was true and it would work out fine but for 'invisible clauses'! The legal ones were easy to work with, but those that he enforced by illegal enforcers were far more binding to Uche. One of them meant that the coop owned him! Being owned by the coop meant doing whatever Tutu Babe bid you to do, or face death! The greed that Uche chose meant betraying the one person who was always good to him. That person was the only one that Tutu Babe feared.

Now here Uche was, out on the deck of the tenth floor of Tutu Babe's apartment building for top-level workers! The big man took care of all his needs, because he was a Special Projects Personnel. He was the one in whom Babe was banking billions of his dollars. He did not need to go outside for anything. Babe did not recommend it, especially for a first-class scientist like Uche! If he thought he had a choice in the matter, he would be dead wrong! His life belonged to Babe! Not to anyone else, including his family! Sure, the boss made accommodations for family too. On his off periods, he could spend as much time as he liked with them. The rest of his life he would dedicate to work within the ten story confines established by Tutu Babe. The billionaire paid for it all!

It was a warm Boston September in Cambridge. Uche sweated profusely from the heat, but he was trembling too!

Both men sat in the blazing sun on wooden chairs facing each other. Tutu Babe was sweating also, but he did not mind at all. "Jeezus Uche! Can you believe we'd get it so hot in September?"

Uche swallowed hard and made to wipe his brow with the back of his hand before remembering that he was strapped to the chair. "N-No s-sir…I didn't know."

Babe pulled out a Cuban flavored electronic cigar and sparked up. He eyed Uche affectionately, nodding his head as if approving the scientist's insignificant words. Hairy Dale, who was standing behind the big man's chair, guffawed and the boss stuck out a hand as a hint to the unrefined subordinate. "Was there a joke asshole? Fuck off with your inappropriate behavior! I'm talking to an intelligent nigger here! Show some respect!"

The three men that were standing beside Hairy Dale stepped away to muffle their own laughter. Hairy Dale did not move. He avoided Uche's eyes. Tutu Babe refocused on the scientist. "I apologize for that asshole's scornful behavior! It's just because you're a *nigger*. You're *black* you know. Or didn't you realize?"

Uche saw no need to respond. They attached his own device to his chair! It was a lie detector, set up on a pulley belt. They turned his back to the edge of the building, where the device rested fifteen feet away. Every lie he told would pull him three feet closer to the ten-storey fall! It systematically allowed him five lies and no more!

"Why do you insult me Uche? Is it because I'm no nigger?"

"No sir…I didn't…What did I do?"

"I asked you a question and you ignored me!"

"*Yes*. I know I'm black."

"I'm glad! Be the best of whatever you are! If you're a nigger, be the best nigger there ever was…like Hip-Hop niggers and black Obama! You agree?"

"I'm not sure what you mean."

"Do the things that niggers should do Uche! Own the 'fuck myself mentality'! Niggers are not allowed to have compassion on niggers!"

"I'm sorry. That's prejudicial and I cannot say I agree."

"Holy *fuck*! You truly think you niggers are human! The pulley didn't move!" The big man nodded with approval, "Even in a half-human race comes one with integrity! Admirable going Mr. Samuel Uche!"

The big man took another draw at his cigar and swept his hand in a bid to urge Uche, "Go ahead and say 'thank you'!"

"Thank you sir."

"Now, let's get to some more pressing affairs! Did you realize that all those dead niggers resurrected on the Ivory Coast?"

"I realize that they did not die sir."

"Why didn't they die Uche? Was it because they were niggers like you?"

"Please let's go back over our agreement Mr. Babe."

Tutu Babe removed the cigar from his mouth in utter shock. "The nigger tells me how to run my show! Who would believe it?"

He could see the tension on Uche's face. The scientist was about to apologize when the big man conceded, "Alright! We'll oblige you! Let's talk about it! We will also talk about what I spent hundreds of millions of dollars for you to build!"

"Thank you sir. The agreement was that I would build this weapon to be used solely at the discretion of the United States Government."

"That hasn't changed! Has it in any way?"

"I believe so. We were using the weapon against people who were no threat at all to the US..."

"Fuck the gobbledygook Uche! We were merely testing the equipment! It's not like we're gonna test it every day!"

"You told me that we would test it on animals!"

"I changed my mind because you tried to trick me! If it's meant to kill humans, you only know it works when it kills humans. Not when it kills monkeys or stuns people!"

"I told you that whatever I simulate is one hundred percent accurate sir!"

"Therefore, what you make must also be a hundred percent accurate! Why didn't it work? Was it because the people were African niggers like you?"

"No sir!"

"If you went to Poland, would it work over there?"

"It would depend on the same factors sir!"

"What are those factors?"

"For one, how the machine is calibrated."

"Huh? Give me a second reason!"

"If it involves human lives sir!"

"Mother fucker! We have been using you shits as guinea pigs for over a hundred years! Now you come to tell me about 'lives being lost'? Would you like me to try the machine on Caucasians in first world countries?"

"No sir!"

"Do you think the machine would suddenly work?"

"No sir!"

"What if we went to Jamaica?"

"No difference sir!"

Tutu Babe breathed in deeply. Then a broad grin came across his face. "Can you calibrate the machine so we could try it out in Kingston Jamaica tomorrow?"

Uche froze, trying to figure out what to say. Finally, he offered, "I don't know sir!"

The belt pulled three feet backwards. He screamed unwittingly. Tutu Babe chuckled. "Do you believe that the machine is a hundred percent capable?"

Uche did not answer.

"Come on Uche! Three feet for every unanswered question! Did the machine malfunction or did you tamper with it?"

Uche swallowed hard, "Sir, please, my family don't know where I am!"

"Do *you* know where those niggers are? They will find you soon if you do not oblige me! Then I will find them! Hairy Dale loves to fuck nigger bitches in anger!"

"Sir, I adjusted the thing!"

"Do you know how to fix it?"

"Yes."

Tutu Babe turned to Hairy Dale, "OK! Bring it here!"

Within a minute, four men came pulling the amplifier to Uche. One stepped up and released the scientist. Uche reached forward and applied the official code again.

"That's it?" Tutu Babe asked skeptically.

"That's it," Uche replied defeated.

"Can I trust you?"

"You already know the answer Mr. Babe! I only wish I was never born!"

"Don't worry about that too much Mr. Uche!"

Babe turned to face his men beaming. "This nigger gave me a hard time, but in the end, I feel so excited that

I must reward him! Henceforth, his children will be spared!"

The big man indicated to the men to strap the scientist back into the chair. "What?" Uche asked in shock.

"Don't worry!" Babe assured him, "You have four lies to go and I only have one question!"

"What is it sir?"

"Are you sure the machine is now ready to work?"

"Yes. It is ready sir!"

Uche passed with excellence! The chair did not move! Babe breathed a sigh and stood up triumphantly. Uche watched, waiting on the orders for his release. Then Tutu Babe spoke. "Take the pulley belt off that chair!"

The men obliged. "Good!" Tutu applauded, "Now untie him and throw him off head-first! I don't want this nigger to cost me anymore! Not even a fucking chair!"

He rose and walked to the stairs, heading back into the building and savoring his cigar. The scientist's desperate pleas were music to his ears. He stopped and turned around to see them let him go. Those men really enjoyed killing people…especially 'ethnics'.

As Hairy Dale caught up with him he offered, "When I said I would spare his family it was to reward him and nothing else! All Uche deserved was to die feeling they would be safe. You can go and take turns with those nigger hoes! When you are finished, get rid of them! I don't want them on my pocket anymore! Ansell at info will give you their keys and apartment numbers!"

21

June 4, 2026...

She was up in bed watching her recording of Gyasi's work when he knocked. The god man had a special knock for her, and she practiced it and used it for him in return. She smiled and rose to let him inside. "Hello! You have not abandoned me then! I've been wondering for days!"

Aduna stepped inside the room. "I was busy. It becomes hard to think. There are things that I must get perfectly right the first time!"

"I understand. So, are you here to see me, or are you here on project business?"

"I am always here for both! You, Dianne, Angel and I...we are a family, but our family is the project too. If I forget to mention anyone it is deliberate!"

She shook her head and smiled in a cheeky way, "Oh there is no catching you on points! Is there?"

"No my daughter. I make the best points possible!"

"Alright then! So what do you want from me today?"

"I want you to send in a report to your work place."

"What? Are you crazy? It's been over a year! I am dead! Dead people don't work in society! Besides, even if I was alive, they would terminate me by now."

"You will be the best news for them! In fact, I would rather you talk to Boston Day. They are even bigger!"

Anne turned her back to him, trying to picture the stir it would make. It would be the absolute hit in media for sure and she would be much more than famous! The masses would want to know where she was but that would be a manageable challenge. She would have to bring something news worthy enough to support the common interest when the first wave of curiosity wore off. Anne breathed in deeply. "Father, explain this to me!"

"I want you to keep all your channels open. Maybe one day you will have to go back to being Marilyn Trooper."

"Why and how?"

"Forget about why. I said 'maybe'. How? Keep your channels open. The story is that you are here against your will. The sleuth that you are, we found you snooping around our facilities. We would not let you leave and compromise our project. We were not willing to kill you either, so we held you here against your will. For almost a year, you tried to convince us to let you film and explain the project to the world."

"And now you're conceding to my plea? I thought Aduna Gyasi did not lie!"

"I do not lie, I 'strategize'."

She smiled ironically. "'Strategizing' sounds an awful lot like lying I must say!"

"I know it does my daughter!"

She breathed in deeply, "Alright! I'll 'strategize' my butt off for you! At least my father thought me how to be disingenuous!"

Aduna chuckled unwittingly. "You will send a weekly report to the media explaining things that we agree to share with you. Say we treat you OK. We will let them see that you are being restrained. Say we will allow you to leave after the birth of the cyber-babies, when the lab will disappear."

"Then they'll pull out all the stops to find me and destroy everything you built!"

"So let them come! I want them to do just that!"

She shook her head confused. "Oh Father! Can any man understand you?"

Marilyn Trooper came back from the dead that afternoon. She sent the video file to Tony Patterson, chief editor of the Boston Day. He watched her in awe, for even in her pitiful state, she was sending him a lifeline! Faith Defender attacks were prevalent, but they were like stale news. A two-hour recording of East Lab operations was just what he needed! He would be the man! She even filmed Aduna making insertions for the cyber-babies! It included charts showing how they would select the prototype for the Arpago. Marilyn promised to send him videos every week. He only had to welcome her as a freelance reporter for Boston Day! That part was easy! She could get anything she wanted! All she needed was to tell them where to send her money! He could see the headline already! *'Missing Star Reporter Marilyn Trooper is Alive!'*

As a mere formality, he sent the recording to the experts to prove its authenticity. It took over three hours to verify her media before releasing it to the public! The entire world stayed tuned to Boston Day News, and the Police immediately reopened the case. They would find Aduna Gyasi, and they would do so before the Faith Defenders did!

On June 5, Charley Boy and Halloway were having coffee at Sheryl's when the news broke. They were not watching it. Sheryl went to them, rapped on the counter and announced, "Your gut feelings are real Halloway! Look what's on TV!"

The cops watched the report in a astonishment!

"Many months have passed since the East Lab security caught me scouting around their barriers. To me, it seems like a decade! For almost a year, I besieged Aduna Gyasi to let me report on his project. I vowed not to try to give up his location but he refused consistently. I am not sure why he has reconsidered and he is not telling me either. I can only speculate. I believe he realizes that the exposure will help him regain his credibility. Whatever we think of him, he believes he is doing a greater good. It would serve his cause to be accepted. He has me shackled like a slave, but he wants me to send a fair report to the world on his behalf! It is hard not to question an abductor's integrity!"

"I want my freedom! Dr. Gyasi thinks that it is too risky to let me go before the project ends! Maybe I was foolish to come out here but I am an investigative reporter. I have a need to know! This could have cost me my life and it still can! However, the graphic footage that I have obtained is good reward for my sacrifice!"

"In the eyes of the people, Aduna Gyasi is the antichrist! His work seems to prove it. Now he asks the world to understand that he means well. If you believe that, you might well believe he means well to have me cuffed like a prisoner too! I hope someone will rescue me soon and bring the criminals to justice!"

They heard Aduna Gyasi's voice in the background. He did not like the tone of her presentation at that point.

Someone put a hand across the screen and turned off the video.

Charley stole a glance at Halloway. The man of rock had his big palm around the beer jug, squeezing, and the youngster feared that it would break. "Charley Boy. Do you know what won't go away until I do it?"

"What?"

"My need to kill Aduna Gyasi!"

"You know you can't do that, right?"

"Yuh! But I'll do it anyway!"

"*Why?*"

"You know why kid!"

"Because you don't give a shit?"

"Right!" he confirmed and then added, "Marilyn Trooper is a brave woman!"

....

By the end of June 2026, seven cyber-babies were born in the East Lab hideout. Aduna Gyasi kept a sober face, watching his people celebrate. Carter patted his shoulder, "This is it man! A lifetime of hard work finally bore fruit!"

"Let us make sure that it continues on the right track."

Carter paused speculatively and then said, "You're right! We don't know who to trust! We got the government, the media, Tutu Babe and the Faith Defenders to worry about. Anyone can attack at any time!"

Gyasi did not answer. Carter patted his shoulder again and walked off.

"Father, I see you don't have your happy face on!"

He turned to face Anne. She looked as sober as he was. "Come my child. Let us go to your room!"

The god-creator walked off without looking to see if she was coming. Marilyn struggled to keep up with him. She caught his elbow as they got to the door. "What's up with the rush? You miss talking to me?"

"I have things I better tell you now."

"You mean *'things that you shoulda told me but you couldn't't'*?"

"Yes my child. This is the day to tell you more!"

He was waiting for her to open the door. She pulled her card and swiped. "There you go prodigal father!"

Aduna went inside and sat down on the chair that was by her bedside. "I see you are doing fine without him," he tested.

"Without *Angel*? I'm not! He connects me quite often!"

She sat on the side of her bed and eye-poked him. Gyasi fidgeted. "My daughter, before you were born, I put Anne Eve's life in danger! You lived many years as Marilyn Trooper, the alien to you."

"I understand that sacrifice Father. Angel told me and I'm proud! Mother explained what happened to Lisa. It scares me, but you must know what to not do this time around."

"I knew what I should not do then and I know what I must do now! I am never wrong! You are a hundred percent safe in the hands of Aduna Gyasi. So was she! Lisa's death was no accident. She was deliberately killed from inside the project!"

"How?"

He ignored her. "Remember I told you not to answer questions about Angel! You do not even know you have a 'twin'! The backbiters regard Angel as a failure. That is my official position as far as they know. You were safe as Marilyn Trooper. The seven prototypes we delivered this month are seven sacrifices! If they live for more than six months, you will die. If they die, you will live!"

"But they're the project! If they die the project will die too!"

"If they die they die! You are the Eve of future earth!"

"Well, I can't die, and they can't either!"

Aduna sighed, "My beloved child! I wish you could understand the pain I..."

"Your pain of heart because of your love for us? I felt it a few times. Angel gave it to me!" She caught his eyes and he could see her tears start to well. "Father, I said many awful things to you, because I didn't know! I was one of them evil *humans*."

"All is fine now. This day you will prepare to carry on for our kind without me!"

"What are you trying to say?" she asked nervously.

"Now you have Angel. You said he spoke about my shortcomings. What did he say?"

"I lied Father. Angel told me to listen to you and learn as much as I can! He said that if I cannot be your daughter I cannot serve my purpose!"

"Oh?" Gyasi croaked, sounding surprised and choked up at the same time, "Then it is time to be my daughter! Starting from today, you will be Great Gyasi's Great Daughter creating Great Gyasi's Greater Son! You will finish the struggle on your own. You will never be alone!"

"Why would that be necessary when you're here?"

"My love, you are the most fortunate being! You are one person in two different places and sexes. I cannot be as great as that! Angel and you are more than a twin. You are one soul. He is merely an extension of your brain and creative ability. I wanted to create the omnipotent factor. This is the beginning of it. The result will be your child."

She was clear and confused at the same time. All these things she experienced, but preconceived notions of norm were compelling her to not believe. Now she heard it from the god-creator himself. She nodded. "Father, sometimes I know and then I don't!"

He smiled with understanding. Then she asserted stubbornly, "I will work and serve you! You will preside over the project!"

Aduna Gyasi had a faraway look on his face. Eve did not like it! Neither did she like the indifferent answer he gave. "Anne Eve, in their belligerent minds, I am the project. Nothing happens without me. I transferred my successes to you. If I do not die, the project will fail!"

"Why?" she screamed, jumping to her feet as if the bed shocked her. "Who said anything about *dying*?"

He did not answer her. He just said, "They will come to destroy us. When they come, you will be our prisoner. Whoever betrays us, will not betray you."

"So we'll move again, and hide before they come!" she offered, hoping that he would consider her words. Gyasi ignored her. His face reflected the futility of her efforts.

Anne sat down. Her face set like stone, tears falling down her cheeks...but she was Gyasi's rock, and with all the hate for those evil ones, she listened. "My daughter, I did not give you a friend among people. I created your friend. Doo-Bug is your legacy. Tell him whatever you cannot tell humans!" He paused to catch his breath and make sure she heard him right. "Now, listen! All this celebration is for naught! There are two projects! The real one and the 'other one'. This here is the 'other one'!"

"So where's the real one?"

"It is programmed inside you."

"But how will I know?"

"You will, in good time, my androgynous child!"

"Am I androgynous now too?"

"In the eyes of a god-creator we all may be."

She shook her head confused, "Father, I don't understand why you work with people you can't trust. It's killing me!"

"On the contrary, it is what saves you. They are close enough to see how harmless you are to their game. Their witnesses have first hand view of all that transpires. When they destroy all we have, they will stop looking to destroy."

Marilyn got the idea. "But what they think they know will not be the right thing, and they will destroy it!"

"Precisely!"

She sat up and composed herself, becoming the rock again. "They know that all wisdom is in Aduna Gyasi and no one else. If Aduna Gyasi dies, the project dies with him."

"As long as his handiwork is first destroyed."

"Therefore, if they destroy his work in this lab, even Eve would be insignificant."

"Right!"

Gyasi rose to leave without saying another word. She reached forward and grabbed his hand. "Wait Father! I was not done!" He stopped and she continued. "So why can't you fake your death the way you faked ours?"

The little man shook his head stubbornly and walked off saying, "Someone will come to you when the time is right! Someone you will be happy to meet!"

"But I don't want to meet anyone!"

Her heart sank inside of her. She was left to reflect on what losing the father she never had would mean to her.

Humans were evil! That alone justified Aduna Gyasi's cause. They would not evolve! They deserved destruction! They were nothing but bloodthirsty vampires! Once upon a time, she was one of them. She was lost and they deceived her!

Anne remembered how they almost killed her in the institution. Many times, they compelled her to eat dead things! After the third time, they proved that she was 'allergic'. She was lucky! Her foster parents were strict vegetarians, although they killed the vegetables before eating them.

Suddenly there was a knock on the door, which startled her. "Hello?"

"It's Doo-Bug Miss Gyasi!"

Anne's eyes lit up. She just loved the robot! It was always good help and good comfort. "I'm coming Doo-Bug!"

Anne got up and opened the door. "Come right in my dear!" she invited sweetly.

"Sorry My Lady!" Doo-Bug replied respectfully, "As much as I love sharing warm moments with you inside your room, I'm here on business! I need you to follow me!"

Its comment was perplexing to Anne and the robot enjoyed it. She caught her bearings and responded with a regretful expression on her face. "Oh? OK..."

The robot headed off and she hopped out of the room behind it. "What now?"

"Top secret! You know the ones like 'if I tell you I will have to kill you'?"

Anne paused and leaned her head side-ways to think. "Oh my goodness! *Yes*!" she exclaimed, whispering with mocked hysteria.

"Well it's one of those!"

She froze, wondering what to believe. Doo-Bug stopped to wait for her and she asked, "Are we talking *seriously*?"

Doo-Bug put on his macho voice. "Lady, I'm a secret agent! I never have time for anything but serious business!"

"If you say so Doo-Bug!"

"Good! The only other person who knows where we're going now is Aduna Gyasi. Watch him carefully and tell me if he gives up the secret!"

Marilyn giggled, "You think he will?"

"Not willingly but under pressure, the weaker ones scream like little girls!"

"I see. I hope you trust *me*!"

"That's why you've been recruited my dear!"

She followed him to the basement wondering if she was going to see some weird science that she was not ready to embrace. The robot produced a remote and a corner section of the floorboards rose. There was a basement below the basement! Moved by curiosity, she hurried along, down the stairs and discovered even more. The basement below the basement was actually a lab underneath the basement!

"Who'd ever figure this out?" she muttered breathlessly.

Doo-Bug pointed the remote at the wall and it moved away to expose the incubator. The huge apparatus turned automatically to stand on its end and then the cover slipped away. Aduna Gyasi was inside! He looked like a zombie!

"Father!" she screamed, petrified!

She charged for the body. Doo-Bug grabbed her by the hand. "It's OK Miss Anne! It's not him! It's like planted meat!"

She remembered that her father had just spoken to her and weighed the possibilities. Her heart began to calm down and her breathing steadied. Finally, she spoke, sounding relieved, "I see! This thing is going to die when the adversaries come! Not my dad!"

"Something like that."

"It's his flesh, blood and bone – I get that. But, how will it work?"

Doo-Bug pointed the remote. The thing opened its eyes and walked out into the room. "It's a machine made of flesh and bone. We can run it on remote, or we can program a routine into it. It's a super-computer. Fundamentally, that is what human beings are. But, you're a Gyasi...you should know these things!"

"Every family has a black sheep Doo-Bug!" she quipped dejectedly.

"Hah!" Doo-Bug cracked with mocked insolence, "You're telling me! You're dumb to the core, but no one cares! They make you the project master!"

Anne glared at the robot, surprised. "Doo-Bug!"

"Yes Miss Gyasi?"

She shook her head with resignation and let it slide. "One thing I don't get though...when Father disappears, what do I do? Where do I go?"

"You go back to being Marilyn Trooper."

"*How*?"

"You already are! Don't you see? Doo-Bug must tell you *everything*! Just because he's the brain in this outfit, he must work overtime all day every day!"

Anne laughed. "Oh I love smart asses!"

"I know My Lady. I'm an asshole to please you!"

22

Weasel cupped the phone to his ear and lowered his voice to a whisper, although he was the only one in the room. "But Mr. Babe sir, Dr. Carter never lets you know everything. He withholds information that is as valuable as what he tells you!"

"How could that be Weasel? He gave me all the information to destroy the project."

"Yes, in his own mind, but you should be the judge of that! The project will continue regardless, unless you know what I know, which Dr. Carter is hiding!"

"OK Weasel, you have my attention! What information do you have for us?"

Weasel paused to think. His heart was racing and he wanted to figure on the perfect approach.

"Weasel? You still there chum?"

"Er...sure. Yes Mr. Sir...er Mr. Babe. I'm here!"

"Good! Tell me what will keep the project going if Aduna Gyasi 'takes a walk'."

"Sir, I really want to tell you! This kind of information cost a lot though."

"It does?"

"Yes sir!"

"And what's a lot!"

"Like, say…twenty!"

"Twenty thousand?" Tutu Babe muffled a laugh, "That's huge...but I will pay!"

Weasel did not answer. His heart was beating fast and his head spinning. "Weasel? Where are you? You don't want the money anymore?"

"No sir...I mean 'yes sir', but there's a little problem here."

"What is it?"

"I w-was not talking twenty thousand. I meant a half of what you're paying Dr. Carter...like twenty million..."

The line went dead for a while, and Weasel thought the big man hung up. He was about to hang up too when the voice replied, "I will pay you the money for sure! Carter James will receive his own disloyal currency!"

Weasel heard the man. He even deduced that it could mean the death of Carter, but he was too high to care. He was going to make twenty million dollars, just for talking all he knew! Something Dr. Carter could not do too well!

"Good sir! Tell me when and where! I'll be there!"

"I'll contact you Weasel!"

Tutu Babe hung up. Doo-Bug heard the click on its line-in system. The god-creator wired all calls from project staff to ulterior entities through it since the day that they retreated to the East Lab hide out. That proved to be a smart move.

"Poor me!" Doo-Bug lamented sarcastically, "They have me nosing into people's business! I'm tired of listening to private farts, especially when they stink to hell!"

Weasel was sitting on the side of his bed beaming when Doo-Bug knocked on his door. "Doo-Bug, I know it's you! Come in you robotic asshole!"

The robot pushed the door and entered. "Since Mr. Gyasi patched you up you have become Mr. Smart Crack!"

"Look who's talking!"

"I'm talking, which means someone's making sense!"

"Don't spoil my day! What do you want with me now?"

"I wanted to show you something to blow your mind, but you're being an asshole!"

"What're you talking a..?"

"Escape routes! In the building! No one can attack the project and capture Dr. Gyasi alive!"

"That I gotta see!"

"Then just follow me!"

Doo-Bug led him to the corridor and opened the secret portal leading to the unknown lower basement. The slender man watched in astonishment. "This is awesome!"

"You haven't seen anything yet! There's a bigger lab underneath here!"

"You don't say!"

"I dare say!"

Doo-Bug stretched its hand to the wall, bringing down the incubator and exposing the Aduna look-alike. Weasel stiffened in panic, and Doo-Bug had to assure him, "It's not Mr.Gyasi! It's his look-alike!"

"Oh...I know," Weasel lied to make it appear as if he was only kidding around.

"I could tell you knew!" Doo-Bug offered, returning the favor. "And the tunnel is over there. Let me show you!"

"That I wanna see!"

They turned and Doo-Bug walked up beside him, palming the small of his back to urge him forward. Weasel felt a sudden chill in his breast. His body became lethargic but he pushed on, eager to see what was ahead. He noticed a drag on his feet, as if something was pulling him back.

"Hey Weasel!" Doo-Bug chirped, "Guess what?"

"What?"

"I stabbed you in the back and you don't even have a clue!"

Weasel smiled his ridicule, but he looked down anyway. There it was! The shining blade protruded from below his ribcage! "What the *fuck*!" he screeched hysterically as his knees began to buckle. Doo-Bug eased him down and pulled the blade out of his spine. The robot stood over Weasel, watching the skinny man die. "Fool, you took so fucking long to know! I knew you would die dumb!"

It walked off, stopped, leaned its head and pulled invisible beard to imitate Gyasi. "All this cleanup work! Always mine to do! It is because of these backstabbing humans! The original prototypes are old farts! We must make better people than these already!"

....

Carter James rested his hand on the telephone. He thought long and hard before lifting the receiver. The lanky biologist had to play his hands well. He would not give Tutu Babe another opt out excuse when it was time to pay! At the same time, he was not going to make the Gyasi's suffer unnecessarily. If it was not for Carter, Anne Eve Gyasi would have died before she could walk, and her vegetable twin with her!

Gyasi was a good friend. There was no reason to hurt his children. All Tutu Babe cared about was to stop the project, and that is what he was paying Carter to do. Anne Eve would be no threat without the prototypes. He would have them destroyed. If Babe knew those children were alive, Carter would be dead! He lied to prove his friendship to Aduna. Since he lied twenty-nine years ago, he had to keep lying now. That was the only way to save his skin. Anne and Dianne were not only Gyasi's secret. They were his secret too! It was a smart move for Gyasi to prepare Anne to become Marilyn again. That would be better for everyone! He watched her news reports from captivity every Friday night and enjoyed the genius of it. The public enjoyed it too. It was better than Hollywood!

Carter was the hero who gave planted meat to the world. Not much kudos came his way, but the world was eating it! Babe received all the money and the glory. Carter wanted his own! Gyasi's friendship was not enough! The wizard was too damned stubborn! Now Carter had to sacrifice the god-creator. Tutu requested it but he would have to do it himself! Carter would only give him their location and the project information.

Although money made little sense to him, Aduna Gyasi had done well financially. Carter wanted to retire worthy of his hard years of labor! This was his opportunity!

Had Gyasi agreed to Babe's proposals, they would all be billionaires! The god-creator insisted that Babe would never pay over that much. Aduna referred to the planted meat contract as an example. Everyone, even the lawyers, agreed that the fault was on Carter! He did not negotiate properly. This time, all he had to do was work on a better contract. Aduna swore that it would not work with Tutu Babe at the helm.

Carter lifted the receiver and dialed the number. Tutu Babe responded quickly as usual, "Hello partner!"

"Hello boss! I have great news for you!"

"Are they all delivered?"

"All seven!"

Tutu Babe hissed, disappointed. "Aduna's stubborn behavior will cost his life! There's too much wisdom in him to lose! What choice do we have? Can you gain access to his secret documents?"

"Secret documents? Gyasi is a god on earth! Everything is in his head!"

The big man paused. "I see. Keep me abreast of what's happening. We just took on a new Faith Defender general! I need the information! We need to see the job done!"

"You're going to act now?"

"Maybe or maybe not! We have months to go, judging by what you say. Right?"

"Yes! They won't be useable before March next year. That won't stop you from taking them out any time before."

"It'll be before March. Don't cause us to screw-up!"

"Have I ever?"

The big man paused. "How would I know?"

Before Carter could respond, he added, "Send me the location map!" Then he hung up the phone.

Carter looked puzzled. Was Babe insinuating that he did not trust him? That was not a good sign! Maybe the big man was having a bad day. Carter would finish what he started regardless. He pulled up the files on his computer and sent them off. Tutu Babe would be waiting, strumming the edge of his desk impatiently. Carter knew him that well!

23

December 30. 2:45 PM.

The old Mustang veered left off Boylston Street and headed into the town of Dorchester. Halloway swerved suddenly, cutting off a Honda Accord in a bid to stay on the trail. The driver honked her disgust and Halloway's jaws tightened with anxiety, hoping that she would stop already! He pulled back a bit, to make sure the subject did not take notice. They were following this driver from as far as Medford near Somerville, and for no good reason except that Halloway had one of his famous gut feelings. The super cop followed these feelings like religion. They closed the Marilyn Trooper case more than two months before but he still believed that she was alive. After more than a year, his *gut* had lost all its disciples. Now the Marilyn Trooper videos were exposed! Case reopened! They needed to find the antichrist and his East Lab to find her!

The cops were tailing Big Daddy Drake Shepherd. He was a real tough nut of the irredeemable kind. Drake rejoined the Faith Defenders two years before, fresh out of prison as usual. It was like the greatest achievement for him...to walk around calling himself a Christian Soldier and imagining that he was doing something useful for humanity. Some believed he only joined for the opportunity to kill and get away with it! Many suspected that he was murdering majority Christians who spoke up against the Faith Defender crusades. He killed as many of those as he did pregnant and suspected women.

Big Daddy Drake loved the smell of blood. The Faith Defenders called him 'the Lieutenant'. Another name he was very proud to carry.

That day when they saw him in Medford, Halloway took interest. The Lieutenant parked across from them in a Burger Joe's lot. Halloway waited for Charley to return with the hamburgers. Big Drake did not see the cop watching. He was talking to someone on the phone and he was too happy for Halloway's liking. Nothing that did not involve death and mayhem could make the Lieutenant that enthusiastic! They removed him from the streets hoping it was for good, but after just four years in the slammer, here he was! Detoxifying evil people never helped! That only worked for metal heads. People like the Lieutenant had to be taken out! It was as simple as that!

Big Drake was driving and talking on the phone. Alfredo eyed him, anxiously listening to every word.

"I'm heading into Dorchester now, doing the round-up. I got Pete and Alfredo with me. The rest of the boys are set to tag along. We'll get it good. I've got the numbers!"

"Wonderful!" Tutu Babe chirped on the other line. "Whatever you do, you must take out the brain!"

"That I know Mr. Babe! That I know!"

"I want all of those fetuses and young ones to count and check them myself!"

"You will get them all!"

"Great! And you must remember! That thing we have for Carter James. He must get it before you make the attack. We don't want him on the compound tomorrow. He'll book a room in the town. I'll brief you, and you can go straight in and talk to him."

"Righty! I'll go as soon as you're ready and see our very useful partner!"

"He's useful indeed! Although he's a fucking rat! What he's doing to Gyasi...how can you do that to a friend?"

The Lieutenant laughed. "Maybe we're better friends to him than the antichrist guy!"

"Maybe! I'll leave you now to do your thing! Keep in touch!"

"Righty!"

They hit Washington Street at an intersection. A Honda Odyssey van turned into their lane and fell in behind the Mustang. It was full with passengers. Halloway turned his right eye to the rear-view, on a gut feeling as usual. Two cars had taken space behind him! He was too close to the Lieutenant for this! The cop pulled over into a Jamaica Restaurant parking space. "Why are we stopping?" Charley asked puzzled, "He'll lose us!"

"We're too close to the lead Charley Boy!"

"What?"

The cars passed by him and took up their places behind their leader.

"You see it kid? It's a motorcade! And other cars are still falling in!"

Charley spun his head in all directions, his brain fighting to process the information. Then he said nervously, "Fuck yeah! There's got to be at least forty persons on this trip! This shit is too big for two of us! We need to alert HQ!"

"Yuh...but we don't even know where we're going. How's that going to help?"

"We need strong backup regardless. If anything goes down, this army will cream us!"

"We're riding on assumption. I'll call and alert HQ, and we'll follow this gang, see what they're up to and take it from there!"

"Whatever it is, this one is big! Maybe they found the East Lab hideout!"

At 5:45, they were still tailing the motorcade. Halloway's hand felt numb on the steering wheel. Traffic was thick on the highway and he was doing his best to stay in line without being noticed. He turned on the headlights and shifted uneasily. Charley pushed back his seat and stretched his legs. "Just where the heck are we going? When will we ever get there?"

"Good question!" his partner shot back wearily. "I don't fucking know!"

At 6:50 PM, they pulled off the highway and headed to Shrewsbury MA. Then the trailing cars separated! This told the detective that they were almost there. The Christian Soldiers did not want to be conspicuous entering the town. Charley radioed in, "Hey Sis! Me again! We'll get back to you on this, but for now, let's see what you can prepare in advance for Shrewsbury MA!"

"Shrewsbury? That's like fifty miles away! You're kidding me!"

"No I'm not!"

"Sam and Alvin took out a tag on you! They're following your car. We have a support team in tow. They talked to our friends from surrounding towns. You're bigger than Holly Wood now! Everyone's watching you! I logged the precincts on to your unit and they're sending out keep close people. OK?"

"OK sis!"

Charley was about to end the call, but Laura stopped him. "Oh! Hold on!"

"What?"

"OK...good news! We want to boost your confidence. They're getting choppers to drop some of our kick-ass boys nearby! How does that sound?"

"I like that almost as much as I like you!"

"Shut up Charley! You don't have to flatter me!"

Charley dropped the receiver and sat with a broad grin on his face. Halloway eyed him. "They're all gone on different routes. We're staying with Big Daddy Drake. See what he's up to!"

Charley did not respond. He had a far off look on his face. He was a brave little cop but he had much less experience than Halloway.

The Lieutenant turned into a dirt track. Halloway stopped and killed the lights. He was wondering whether to continue behind the thug, but Charley said, "Back up to the shadows. He's stopping!"

The youngster hopped out off the car as soon as it stopped. "I'll walk down and see! Don't move a muscle!"

The three men exited the car. They stood on either side of the track. Charley stayed in the shadows, keeping his distance. Maybe they were on a stake out.

The youngster heard footsteps behind him and thought it was Halloway. He pulled to the side and turned around. A chubby boy was heading past him, towards Drake and his crew. "Howdy there kid!"

"Howdy mister?"

The cop pointed down the track. "What's down there?"

"You mean where those three men are standing around the car?"

"Yeah!"

"It's the Parker's Ranch sir!"

"A ranch? Thanks pal!"

"You're welcome sir!"

The boy headed along. Charley lifted his gaze back to the men. They were going back inside the car! He hurriedly retraced his steps to the main road. Halloway was leaning back on his seat, his eyes seemingly closed. Charley smiled, pulled the door open and went to sit beside him. "Could be a stake-out!" he offered.

Halloway drawled back lazily, "So I thought. What's down there?"

"Nothing...just a ranch. Parker's Ranch."

The big cop was about to say something, but he was distracted. The Mustang pulled back on to the street and went in the opposite direction. As it sped by them, they caught a glimpse of the devil himself.

Charley watched the muscle man, Big Daddy Drake, calm as ever, eyes intent and cold as frozen steel. That satanic presence was a bit of a rush to the young cop. He braced himself as if expecting something to happen. The red car tore by, shaking their unit with unbelievable velocity. He swallowed hard, relaxed, and wiped his clammy hands on his knees. Halloway was the other rock. He was unmovable, unfeeling; ready to take on life or death like nothing new was happening. The big super cop waited until they were far ahead before firing up his own engine.

The Mustang drove through the town at 7:00 PM. Drake parked the shiny red car next to a motel and got out. The cops took a turn and drove around the block. Then they came back to park about thirty yards behind their subjects.

"Now we wait again!" Halloway agitated. His stomach began to churn and he remembered the hamburgers. Charley had placed them on the back seat. He was in the same mind as the big cop. The kid turned and reached for them. "I can eat cold hamburgers right now!"

"Yeah! And grab mine too! Where's my coke?"

"Easy big boy! It didn't go anywhere!"

They watched Drake exit the car. He was in no rush at all. The Lieutenant walked up the steps to the front office, dressed in army fatigues...and a real soldier he was too. At over six feet tall, he had muscles like a water buffalo. Daddy Drake was intimidating!

Drake's cohorts remained in the car. Halloway swallowed a mouthful and chuckled.

"*What*?"

"We could be here all night! Maybe we should get more burgers before this town closes down!"

"Who's saying it will?"

"It's a remote town Charley!" Halloway drawled lazily. "It'll be worth it when I kill that serpent being!"

"You sound like one of those Faith Defender people. That's a bad thing!"

"We're on the same side! They only kill too many innocent, plus they're not authorized!"

"Authorized to kill? We're not authorized either!"

"That's not what I meant Charley Boy!"

Carter was waiting for the Lieutenant in room A16. The door flew open for the big man before he could knock. "Got it?"

"Yeah, I got it," Drake assured him nonchalantly.

"Finally Mr. Babe delivers!" Carter hissed in a sarcastic tone. He backed away to let the visitor in. The giant took two steps forward and pulled something from his pocket. Before the tall scientist could say "Come on in!" the Lieutenant dropped his voice to a whisper and offered "Well here it is!"

Carter grinned expectantly. Drake's hand went up in a swift arc and 'swoosh', a neat slit appeared on the scientist's throat. "There you have it, meat man! Wait no more!"

Carter's eyes widened, bulging from their sockets in utter disbelief. He did not feel pain. In fact, he felt nothing at all at first, but then he could not breathe. He panicked, and heard himself gurgling for air...then he looked into the big man's eyes to question. Drake smiled and nodded the answer in his face. The order was confirmed! Carter knew that he was dead. Why was he betrayed? His mind went back to Gyasi and the things the great man tried to teach him about Perpetual Motion.

Halloway saw the big man come back down the steps and head for the Mustang. He did not turn on his own ignition until they were far ahead.

They followed the Mustang for about three or so miles. It pulled over into the parking lot of a restaurant named 'Barbara's'. They waited, feeling like they had not eaten at all that day.

10:30 PM, the business closed. Charley frowned, about to ask where the men were. Then Alfredo came out, followed by the other fellow. Both of them were obviously drunk. The Lieutenant stepped out after them. He was in no rush, just taking his time, calm as usual as if the world had to wait for him.

They drove off, and Halloway picked up the trail again, wondering how long they could do it without being noticed. The Mustang went back to the same motel where they stopped before. This time all three men got out, and it seemed like they meant to rest there for the night. After hanging around for fifteen minutes, the cops left to fuel up and grab whatever snacks they could find at the gas station. They would stay right there in the car and be bored all night!

....

Two Shrewsbury boys came to relieve them by seven in the morning of December 31. No one knew exactly what was going down, but all precincts joined forces to play the hunch. At 8:30, Charley made a call to check in with them. They reported no movements so far. That was understandable. Most likely, Drake would do whatever he was up to under the cover of darkness.

Charley and Halloway checked into a motel, a mile or so down the road. They were dog-tired. Sleep was all they could do now.

At 1:30 PM, Charley jumped from the couch with a start and checked his watch. They had slept for over four hours! He grabbed his phone and checked in with the Shrewsbury boys. "What's happening partner?"

"Nothing pal. Mustang still hasn't moved. HQ got no telephone communication for them on the grid!"

"They're making calls though! That car will move at some point!"

"Yeah! Maybe tonight it will. We can only wait."

"I know..."

Halloway said something and his partner paused, trying to concentrate on two things at the same time. *"Charley!"*

"Gotta go!" Charley said to the Shrewsbury cop. He hung up and eyed Halloway questioningly.

"Get your shit on Charley Boy! That was Sam on the phone! Something's going down at the back of that Parker's Ranch place! HQ intelligence! It seems like they did find Aduna Gyasi's hide out! Loads of suspicious units are heading out there! Sam and Alvin are on their way but we're closer! Let's move!"

Charley fought to get his trousers on. "What a hassle! I need a doggone shower so bad it's killing me!"

Halloway chuckled, "I know! I have a nose!"

Charley eyed him disgustedly and changed the topic. "I don't get it man! The Mustang is still there! Big Daddy Drake hasn't moved!"

"I think I know why! That motherfucker saw us following him last night! He's keeping us in one place while his crew goes to work!"

"Don't we have a couple of surprises for him too?"

"I hope so Charley Boy!"

Halloway went through the door and his partner hurried to catch up. The Parker's Ranch was no more than four miles from Lindsay's, where they stayed. They took a path through the woods that led them to the back of the East Lab hideout.

"*Hey Charley!*" The voice came on the radio. It was the Shrewsbury boys. Halloway grabbed the device, "Yeah pal!"

"*The Mustang is off! It's moving fast, heading for Parker's Ranch!*"

"Those buggers! Don't stay too close, and don't lose them either!"

"The boys in the sky are on it partner!"

Halloway hung up the radio. "All this shit makes me want to retire soon!"

Charley turned off the dirt track to a paved side-road. It was very steep going. Up the hill they went, hoping to take a vantage point outside of the Faith Defenders view. They were alone at that stage. How long it would take the others to arrive was any one's guess.

It was near zero degrees on that frigid Shrewsbury December day. Between scant cloud covers in the sky, the ice-cold sunshine smiled down upon the secluded town. Charley leaned his head against the window, admiring the tall snow tipped pine trees lining the street on either side. Birds sang cheerily, each translating a living joy for its own kind, in its own tongue. The sounds and pronouncements of life in exuberance buoyed the young cop's heart. All was good, except in the human realm. They were not here to join in life's symphony of bliss. God did not make humans that good!

He stole a glance at his partner. Halloway was deep in thoughts. "I might live to retire one day!" Charley offered matter-of-factually. Halloway's lips curled back in a tight grin that seemed painful to do. Charley smiled awkwardly. Then the snow came. It was coming hard and dry, enough to cover the windscreen in a matter of seconds. "God!" Halloway lamented, "The sun is shining out there and it's too cold to snow! Where is this coming from?"

"It's the changing times. There's no 'apocalypse' – just paradigm changes and reversals! Besides, the temperature had risen!"

Halloway shook his head distrustfully, "Who are you now kid? Aduna Gyasi?"

Charley leaned his face against the window again and looked out. It was beautiful out there but they were in no position to care. He sighed. "The hearts of men have waxed cold."

Halloway turned to the youngster, a befuddled look on his face. "What kid?"

"I said the hearts of men have waxed cold!"

Halloway was confused, but before he could say anything, the wheels took a skid. He stiffened, "Easy on the gas kid! This hill is a steep mother!"

Charley bit his lips and concentrated hard. Then they reached the top and he was looking down into the valley. There it was! What no one had been able to find for so long! Aduna Gyasi's East Lab hideout! From where they were, anyone would believe that it was a regular farm...extremely large, with many gigantic storage facilities.

Halloway signaled to Charley. The youngster called in. He talked to Laura and Alvin, while Halloway viewed the periphery with trained eyes. Charley dropped the receiver and stood beside his partner, feeling like a small man in a very big world.

To the east were large open acres, so much that they could not see the boundaries. On the western side, there was the Parker Ranch border. A barbwire tipped fence ran all the way to the front of the properties. On the south side, they saw the entrance. That became the focal point for Halloway's lion eyes. He waited. Suddenly, Charley's jaws dropped. A chill ran down his spine and he fought to calm his nerves. This was it!

A huge lorry broke in through the front gate! Faith Defender activists jumped from the back, dressed in their demonic robes and all carrying assault rifles! Halloway exited the car with his Glock 9mm pistol in hand. He had a Navigation Hand Gun strapped to his left hip, but he could only use it for hostage situations. He half-smiled at the irony. Aduna Gyasi was the inventor of most of their pioneering equipment!

Charley took a deep breath and followed the queue. The youngster had a quizzical look on his face. What could they possibly do here without back up?

Halloway hustled down the slope, walking, slipping and sliding. Charley followed him. He stopped, a good distance from the back of the compound, waiting. "I haven't seen any East Lab staff. I bet they'll charge through the back! We'll buy them time until help arrives."

"OK." Charley replied hoarsely, checking out his little gun and then looking down the slope at their rifles. Halloway waited until the youngster refocused and said, "That boulder over there will be good cover for us! Don't worry kid; you're safer than you think!"

"Sure!"

They moved to the cover and dug in. Charley grinned sarcastically, realizing what their position meant! They could not do much, apart from trying to hold off the zealots until help came! If any resident made it through the back door, the cops would cover that person. That would buy them time to escape into the woods on the east side.

From where they were, they could still see the entrance. Faith Defenders surrounded the buildings, rifles raised, ready to kill. The majority made a beeline for what seemed to be the main quarters, nearest to the cops. He watched a line of men take up positions, just ten or so meters away from them. They stopped, held their positions, and waited.

Charley was about to call Alvin when the Mustang came in at full speed. It screeched to a halt beside the lorry. Daddy Drake and his two companions jumped out and headed for the main building too. As usual, the big 'Lieutenant' took his sweet time about it, as if he was daring the world to go ahead without him.

Drake realized what the fences between the buildings meant. There were those who knew they were farmers and others who knew they were working for Aduna Gyasi's project. There entrances and exits were on different sides, and the roads to these sections were completely different too! These two factions would hardly have a chance to make acquaintances. It was a work of genius.

Big Daddy raised his left palm in the air as he moved forward, keeping the faithful on halt. He had, cupped underneath his right armpit, one of the old Tutu Babe remodeled classics – an M-16 Assault Rifle. Tutu Babe was the master of pirates!

The Lieutenant stopped at the front door, a broad grin on his face. "Seems like we're not welcome here today! All of a sudden, they're closed!"

Alfredo cackled raucously, which irritated the big man. The subordinate caught Drake's eyes and stopped prematurely. Pete looked at his counterpart and shook his head with disdain. Then Big Daddy gave them the signal to drop the door. The two men went ahead of him, firing on rapid. They showed no sign of letting up, and the big man hollered, "Fuckin' stop already fools!"

They seized and eyed him nervously. "Well, open it!"

Pete stepped up with a swagger. He raised a big army boot and kicked hard against the door. It gave resistance and the little man toppled over on his back, rifle in hand. Alfredo snorted. Then he cast an anxious eye at his boss, biting his lips to control the unwanted guffaw. Big Daddy moved swiftly forward and booted Pete away, easy as if he was weightless. "You! Other fool! Go open it!"

Alfredo hurried ahead nervously. He reached out his hand and pushed it open. The big man rewarded him with a thumb up. "Alrighty!"

In one sweep of the hand, Drake ordered his people at the front to move forward with him. Those surrounding the other structures stayed put. If anyone escaped through the back door, he or she would simply run into the bad boys in the rear. Either they would stay and accept their fates or he would flush them out to that same fate!

The three went in casually, rifles in hand, and a line of zealous ones behind them. "Hello the house!" Drake called jovially, "You have visitors!"

Pete butted in, "Now to find those demon babies and that Aduna Gyasi and take 'em out of their misery!"

Drake turned his blood shut demon eyes to the little man and spoke, his crude voice like gravels in river sand. "We kill everything that moves and secure the fetuses! All excluding the reporter. She's on our side and I'll take that reward for her!"

Alfredo opened his mouth to speak, but the big man's eyes intimidated him. This was not what he expected!

They were there to the calling of the Lord, not to murder the innocent...except pregnant ones that they sacrificed for the good of all! He could not help but question, "What does Mr. Babe do with all the fetuses we haul in any way?"

"What if I put a bullet up your nosy ass Alfredo?"

....

Meanwhile…

Tutu Babe was the night's guest on *'A Date with a Billionaire with Jennifer Gooden'*. The host beamed at the super mogul with awe-struck admiration, which went way beyond her moneymaking showmanship.

The big man was looking smart as ever in expensive, silky looking, purple and black neutral cloth material. His clothes were designed as ordinary as possible, but only to fit his freshly sculpted form. That did the trick!

"Pardon me Mr. Babes. You do look *fabulous* tonight!"

The big man accepted the compliment graciously, and then threw modesty out the window in a bid to humor the audience. "Sure I do!"

The admiring audience broke out in appreciative laughter and Babe raised his fist like a soccer player who had just scored a goal. "Well Jennifer, I must admit that you look beautiful yourself! And thanks for giving me a right to zero in!"

Jennifer shrugged involuntarily. She was blushing, and the audience loved Babe's pickup line, so they cheered excitedly. Tutu Babe was a single man, and he was many a single woman's dream, especially since his excessive weight disappeared overnight! The host sobered up and said, "You seem to have a lot of practice at picking up women!"

The audience screamed once more and the big man chuckled. That sufficed for an answer. Jennifer spoke again. "I notice that everyone still calls you 'the big man', although you are not like…'*big*' anymore!"

"You mean I'm not *fat* anymore! No, I'm not! But I'm still six-foot six…and I'm still the big man in terms of 'the boss'!"

"I get you! I also know that you're well over fifty. How come you look like twenty-five? Does money make people stay young?"

"Yes and no Jennifer! Yes and no! It takes loads of dollars and a secret recipe!"

"What do you mean *'a secret recipe'*?"

"You sure you want to know?" Babe joked in a whispering eerie tone.

"Sure!" she shot back sounding hush-hush.

"I love youthful energy! I just feed on the thing!"

Jennifer blinked at him, a mocked show of ridicule on her face. The audience laughed at Tutu's corny joke and the effect he had on the host.

"So Mr. Babe, you always have some new product coming for the youngsters. Is there one up your sleeve now?"

"Well…" the Billionaire countered, rolling up his left sleeve and inspecting. "Maybe one for the young and the old is up there somewhere!"

"Knowing you, the fact that you found it up your sleeve, *literally*, doesn't shock me!"

Tutu Babe chuckled. The audience laughed.

"Tell us about it then! The suspense is killing us!"

"Don't die on me you'll! I can't eat older folks. I said *'youthful energy'*!"

"You're so funny! Maybe that's your secret!"

"Let's put fun and joke aside now Jennifer. I have a product that will be the bomb! We had it for many years, but as you know, before there was the 'separation' we were spitting out our good stuff too fast to keep up with ourselves."

"You mean before Aduna bailed out on you?"

"I mean before I had to part with him, but I won't go into that one now! Anyway – About this product! You remember up to around 2020, everyone wanted to look like everyone else. Our sense of fashion was about prescribed brands and designs. Suddenly, we seem to have jumped to a more advanced independent mindset. Now we all rather go for the unique look. If you are Caucasian today, that is no unique look. If you are Indian, Chinese or African, what is unique about you? Absolutely nothing! Tutu Babe came up with a solution to fix that little problem. I mean, what if you want to be rainbow-colored. What if you want to be bluish looking like '*Avatar*'? I'll say it now! You don't have to be black anymore! You don't have to be white anymore!

You can be whatever you want to be with our new Nana Tincture! One little drop to the color or color spectrum required – One little bath and 'Abracadabra', you're it!"

"Wow! Mr. Babe! Do you mean that I can just take a bath and turn up with zebra stripes if I want?"

"Exactly!"

"Wow! I mean, *'wow'* again! But…is the process reversible?"

"One little 'neut' shower and it goes away or you change to a new design! You can look different every day of the year or you can look the same forever. Your choice!"

"A *'neut'* shower! I won't even ask about that! Will there be a drawback to this technology though? I mean, with people changing their looks like that, you may not be able to identify anyone!"

"Our rogue scientist did think about it while he was making the product. Now, what I'm going to tell you is scary but it is no secret anymore. People are already comfortable with it; knowing it was a necessary scheme to protect them. Long before 2020, Aduna Gyasi made chips the size of a bacterium. Later on, the authorities tried to use a monopoly on health care to have everyone chipped. They did it in the name of immunization vaccines. It was a limited tactic, especially for execution in some third world countries. Then they commissioned my cooperation to make cyber-bugs and mosquitoes to carry these chips in the same way they carry viruses. By the wisdom of Gyasi, the chips came with biological qualities. They could reproduce and pass on to the next generation of bugs, as any bacteria can. When a bug or mosquito bites you, you are chipped. Why do you think we cannot get rid of annoying bugs and mosquitoes in most of our residential areas worldwide? We can, like we can snap our fingers!"

"In fact," Babe continued, "now I have a product to release that will make every single mosquito go away! Why will I do it now? You are already chipped, and when your new baby is born, it will be born with that cyber-bacterium. So, let's answer your question about drawbacks! The only drawback is in finding Aduna Gyasi and squashing his project! He made all these things, and he cannot be defeated by them! A kid who becomes a zebra to commit a crime, on the other hand, will not get away. There's a grid that can tell us where you were at any given point in your life! This grid is in the private hands of Tutu Babe Cooperation. With the use of a subpoena, the state can obtain information about you. No subpoena will be required for us to give certain information though. For example, we would be obligated to detect who was at the store holding the weapon when that store was robbed."

Jennifer glared at the big man with dropped jaws. Finally, she blurted, "Wait a minute! Did you just say that I'm *chipped*?"

"Don't worry! Your privacy is protected by Federal Law!"

"Is that last comment another one of your jokes?"

Simon Muir sat in the boss's living room, watching the big man on television and grinning. "Brilliant Mr. Babe! Brilliant!"

Albert the butler was sitting across from the scientist. "What's brilliant about that?"

Simon eyed the fellow, watching him as if he was an inferior creature. "Do you think that all Mr. Babe said was just for idle amusement? Everything was perfectly planned. Like the grid talk for example. That was the bomber!"

"The bomber for what?"

"Do you understand the value of privacy? When people get scared of being on the grid it will become easy to sell illegal off-the-grid solutions! Lawyers and doctors – even politicians will buy it! You know how much money we can make for every million people who buy into it worldwide? We have many hood-rats looking for work to do! We could get them out there selling off-the-grid phones and so on!"

24

Anne Eve was in her room working on her Friday report when she heard the barrage of gunfire! They were expecting this, but not in the daytime, especially on New Year's Eve! She should have prepared, but now she could not help questioning everything. The thunder of reality seemed too much for her! The entire building shook and she folded up desperately in bed, plugging her ears with her fingers and trying hard not to scream.

Then the Gyasi genes took over in her! Anne forced herself up on her knees and crawled off the bed heading through the door to her father's room. He wanted her to be the project master, and she was going to take over now!

She met him in the corridor on his way to her. His eyes looked desperately wild, and she felt for him. He was a scientist, not a warrior...although she never knew him as one to fear. Anne grabbed his elbow, pushing him back in the opposite direction. "You're going the wrong way Father! Get out of here! I know what to do!"

"Anne!"

"Shut up and go Father!"

"But I..."

"You're no longer the project master! So here's what I expect you to do! Go through the basement, head into the tunnel and off to my mother! I'm going to make you dead in the eyes of the state and Tutu Babe!"

Gyasi smiled. Although he could not fully understand her, he knew who she was. That was his sweet Anne

Eve! He looked into her eyes and saw Aduna Gyasi in them. It made him agree to obey. The god-creator froze. Then he turned around and started to walk.

"Wait Father! Before you go..."

He stopped and she stepped up behind him. He turned around to find out what she had to say and fell into her embrace. Anne kissed his cheek and said, "Go on now old man!"

Gyasi felt insatiable bliss as he went for the secret portal. She was the closest thing to heaven he knew! The doors were coming down, they had little time, and her heart was overworking! Anne felt like she would suffocate with fear!

She turned to go back to her room. Doo-Bug stood by her doorway with a pair of handcuffs attached to a chain. "Come here slave! You've been naughty, haven't you?"

They hurried inside the room and bolted the door. Almost immediately, there was shuffling outside. The zealots were forcing residents in Anne's direction! Those desperate souls would try to escape through the back door. The Faith Defenders were methodical, checking room by room. They would soon find her!

The gunfire seized for a while. Then it started again, but briefly. Anne thought her heart could not go any faster, but it did. She heard the most horrific screams ever and her mind deciphered. Sporadic gunfire, lots of screaming...screaming stops suddenly! They were killing the residents! That was not expected! "Oh my goodness Doo-Bug! They're killing us! We need a change of plan!"

"That's where you're wrong!" Doo-Bug announced. "You're not one of the 'us'. You're one of the 'them'.

They're killing 'them', and partly because they're pissed off over what the 'them' did to you! Got it...Miss Marilyn Trooper?"

The robot was confident but machines always were! She shivered so bad, it affected her breathing. "Doo-Bug, are you sure? Because...I can't be dead – Right? You know I can't afford to be dead!"

"Very well spoken! Your papa would be proud! Can't die for the project, huh? Don't you worry, Doo-Bug is here! *Jackpot*! You've been chosen to not be dead!"

"Doo-Bug...you must know that I am the project..."

"You must know that I, Agent Doo-Bug at your service...I protect the project!" Doo-Bug saluted, and she found enough courage to see the joke. The voices she heard were of human sacrifices to the project. They were dying to help her live, and she had to accept that and be strong anyway! Anne Eve giggled, and believed in the machine.

…………………………

More shots were fired. The building shook as if it was resting on the subway station. They could hear desperate spine chilling screams from the residents on the lab section! "Oh my goodness!" Charley croaked, eying his partner for support. Halloway's face was set in stone. He was not there to have feelings, only to do what he did best. The big cop was cold as a lizard, but in times like these, one would be happy for the devil at his side.

"Good!"

"What?" Charley asked, looking to the entrance. Sam and Alvin had arrived. He saw them exit their vehicle and take cover behind it. Then Charley realized that they could not make much difference at that time and his heart sank back to rock bottom.

One Faith Defender spotted the cops and alerted the others. A fat one at the front of the building went for his telephone, obviously to call Big Daddy Drake. They would not worry about two nosy police officers but Drake had to know everything that went down! The Shrewsbury police could not stop them! They could finish and get out of there before the fools brought in outside help! The Lieutenant would show no mercy! He barked into the receiver. "Kill them, and watch out for their backup boys!"

"I hear you boss!"

Drake was working from the official lab side of the property. He could see everything on the farm side. He realized that those on that section could not see through the fence! It was like a one-way mirror! "Fucking wise guys!"

The shooting and screaming intensified. Then the back door flew open. Nurse Hudson was the first to exit, ushering a group of young women outside. She was in her white uniform. The four women that came out with her were cuddling babies. Were those the infamous cyber beings?

There were about six heavily armed men waiting in line for the escapees. Charley saw one of them slowly turn his rifle nozzle to point in the women's direction. He stiffened unwittingly, wondering what next and waiting for the queue. If they attracted that kind of attention, the riflemen would fry them alive! Halloway did not care for calculative thinking! He raised his pistol and fired! "Get your head down Charley Boy!" he screamed urgently as the bullet tore into the Faith Defender's face.

Before they could turn their heads down, the rifles were beginning to swing on their side of the hill. At the same time, they saw the army at the front open a barrage against their arriving counterparts!

Charley's mouth felt pitch dry in the instant. His heart beat so fast that his body felt weak! "Oh god!" he croaked, "What the fuck are we into?"

Halloway spat out an evergreen leaf that he was chewing like a toothpick. "What a fuckin' difference a day makes!" He grated disgustedly. Then he tasted the juice and spat again.

Charley heard the shots pass over him like rockets. He could literally feel the turbulence as they came in close proximity. The rock was holding firm, but gashes were appearing in it. Sharp chips flew like miniature bullets and the wind blew their powder in his face. There was another rock slightly to his left. Chips were coming off it and tearing into his skin. He dug in, shivering and thinking that he was not going to move...but he had to...or they were going to come up the slope! Moreover, they were going to kill the innocent residence!

Halloway raised his head and fired. Charley heard the target grunt, not able to make it into a scream. He found himself joking, even in the hell he was in. "Do you ever miss Halloway?"

The kid put up his head, fired, and missed his target. Halloway stated, "I try not to...and you should do the same!"

Charley bit his lip and put up his head to fire. His eye caught hope at the front gate. Backup had arrived! They had a real army coming in and the Faith Defenders panicked. Those on the outside began to retreat. He fired, and this time one of the men fell. Then they began to move away, but not before one of them determinedly turned the rifle on the women. The cops tried to take him out, but the hail of fire from his cover forced their heads down. When Halloway peeped from behind the boulder, the victims were dead or writhing on the ground. One survivor rose unsteadily. She was trying to make her way to them, slipping and sliding back down the slope in futility.

"Come on Charley! Let's go!"

Halloway rolled off down the slope, and his partner went with him. The detective grabbed the woman's shoulder and said, "Stay here, and don't move! Lie down until the fire seizes!" Then he looked into her face and his eyes widened in disbelief. Charley frowned, watching him and the woman at the same time. "*Hope?*"

"John Halloway..."

"Why are you..?"

She pointed down into the snow, cutting him off, "That's why! But it's all fucked up! Makes no sense anymore – Does it?"

Halloway looked and saw it. His blood felt cold in his body! The thing was like a blob, but it had a face like human. It writhed on the ground, dying. Its eyes brightened the snow with a bluish-green light.

"*Oh my god!*" Charley boy screamed girlishly. "*Ewe!*"

"Unbelievable!" Halloway marveled, "What kind of alien being is this?"

"It's a boy!" Hope offered cynically. The detective looked down at the thing again. He pointed the nozzle and fired. The bullet ripped into its head and the light slowly faded. "Not anymore Hope!" he croaked disgustedly. Then he walked away from her heading for the back door. Charley tugged at his elbow, "What was that all about partner?"

Halloway was raging mad. "Not now Charley Boy! We got important things to do! I'm going to kill Aduna Gyasi!"

Already, many of the zealots had stopped engaging the security forces, and were surrendering. Daddy Drake and his crew were still inside dealing death to the innocent and the guilty ones alike.

Inside the hidden lab, Aduna Gyasi pulled down the meat copy incubator. He hooked it to a pulley that would haul it upstairs to the official basement. The god-creator worked hard, performing these tasks without conceiving why. It was as if external forces were driving him.

They were unprepared because they let their guards down. They did not expect a daylight attack. It was a big mistake, and a foolish one at that. The little man was running against time. They would soon be coming to the official basement with her, and he would do well to be gone! He ran up the stairs ahead of the empty incubator and pushed the door open. The box rode up on the cable and rested down in front of him. He slid it over to the wall without knowing why he did it. What was happening to him? It seemed that he was living in a state of dementia! Most of his memories were gone!

Aduna shuffled through his coat pocket. The remote! He left it in the secret basement after resting it down to attach the wires. Something told him that carelessness was not his style. Gyasi hurried back to retrieve it.

25

Weasel's cell phone rang, startling Anne. "Calm down My Lady!" Doo-Bug said assuredly, "Everything is coming on fine!"

He looked at the screen of the instrument. It was Tutu Babe. Marilyn turned to Doo-Bug, eye poking him inquisitively. "It's not my call! It's not even my phone!"

"I know! You don't use cell phones. So now you puzzle me!"

"Secret agents always puzzle people my dear!"

"Oh I see...So whose phone is it?"

"I don't even wanna know anymore! I took it off the ground!"

"You found it on the ground?"

"Kinda sorta! Maybe it's just some stupid entrepreneur trying to call up some stupid dead fellow!"

Anne giggled. "Is everything a joke to you Doo-Bug?"

"Everything about humans! Go figure!"

"Go figure on that? We have way more important things to figure now; like how to stay alive for one!"

"You don't worry about how you will stay alive. That's Doo-Bug's job. Do you think Aduna Gyasi is as dumb as you?"

"*Doo-Bug*!"

"Oops! Sorry! Robots are *so* direct!"

The Police had suppressed the outer forces, and the firing was dying down. They were pressing forward now, trying to enter the main building. The Lieutenants men were occupied, keeping them out. Now he had only two cronies to run the aisle. That was good enough though. The lab people did not bear arms.

The three were at room seven, heading closer to Marilyn at thirteen! Drake was taking his precious time, even with the cops trying to gain on them from the outside. That was a testament to the character of Daddy Drake. He was his own man.

Charley Boy and Halloway came in from the back. They were at room twenty-four, heading to thirteen, which was around the bend.

Halloway indicated to his partner. Charley rapped on number twenty-four. "Police! Open up!"

"Do you have a warrant?" a voice called back, shocking them. They thought that everyone had exited the rooms in a bid to escape the Faith Defenders.

Halloway butted in, "We don't, but we're here to save your ass! Are you interested in the offer?"

The resident went silent for a while. Halloway was about to break the door in, but then the lock began to turn. A woman came out. "Detective Halloway! I know your voice!"

"What're you still doing in there?"

"I was at the front door when they came! They circled the building, and there is no escape around the back!"

"Smart reasoning! Now is the time to move! Go through the back door and keep your hands up for the Police. Alright?"

"OK Mr. Halloway! Thank you!" she said and headed off.

Around the corner in the aisle, Drake put his finger on his lips. The two went quiet. He heard something!

Halloway pulled his partner inside the room and half closed the door. Drake held Pete by the scruff of the neck and pushed his head around the corner. Nothing happened. He indicated to Alfredo, who stepped out into the aisle. Then Drake calmly joined his team.

From the back door, Sam and Alvin entered, walking gingerly into their direction. The three men flattened themselves against the door, but not fast enough to miss the eyes of the partners, who hit the ground together and pointed their guns in that general direction. Then they saw the rifle barrels protruding from their places of cover. "Bomboclaat!" Samuel lamented hoarsely, "We raas dead today!"

Alvin was so nervous; he gave up after two useless attempts to talk. Suddenly Alfredo moved from his position, firing as he ducked away from the wall to take cover on the other side. The cops felt the ricochets tearing into their faces. Even if the riflemen kept missing, they could not endure this all day!

"Now I really want to piss!" Alvin admitted sadly.

"So piss!" Sam barked, irritated. "We got bigger things than decency to think about!"

The three men were edging closer to them. Sam raised his gun and took a shot at a piece of Alfredo that was exposed. The bullet scraped by so close that it cut into his shirt. He screamed in shock, and Sam let off a couple speculative ones. The trio edged closer again. Drake said "Pete!" and the little one duck passed and let off a barrage. Alvin screamed. A piece of tile lodged into his shoulder blade.

They were standing by room twenty-four now, and Alfredo backed into it for cover. His head exploded and super Halloway was standing where he should have been!

"Watch out boss!" Pete hollered in horror, turning his gun to fire before the big man could move. A third hand appeared from underneath Halloway's armpit and a gun exploded in it. The bullet caught Pete flush in the face. He fired off a volley into the roof as if to salute his coming death. Drake took a huge dive, back in the opposite direction. He charged for the door, where he had a number of men fighting to ward off the security forces. He needed a couple more riflemen to put down the security people inside. When he returned with his support crew, they saw more than a dozen officers coming in from the back, loaded with heavy-duty equipment.

"What a *bitch*!" he hissed disgustedly, and then ordered the men in defiance, "Shoot 'em!"

The men promptly dropped their weapons and pointed their empty hands to the sky. Daddy Drake looked at his own rifle, thought about firing, and then put it down too. The army had many trained on him. "Alrighty! Alrighty! I'll fuckin' put it down this time!"

....

She could not hold back the tears. The cuff was just a bit too close and Doo-Bug decided to do some robotic art on her. It burnt the cuff marks into both her wrists and ankles. They looked like a whole year of wear. The robot linked the handcuff on to a chain on the metal bedpost, just enough to give her space to move around without being cramped.

The firing seized and she wondered what was happening now. Did the law prevail over an army of Faith Defender warriors? Were the cops even there?

"They're coming My Lady!" Doo-Bug announced.

"How can you tell? I hear nothing!"

"I'm a secret agent, remember?"

Anne inhaled, feeling her heart rate increase again. The Faith Defender activists were not always trustworthy. Occasionally they would just 'feel like' killing everyone!

"Stop fretting My Lady! No one can harm you while I'm here!"

"No one? That's an army!"

"Hah! An army of humans like you! Give me a break!"

There was a knock on the door, "Who's in there! Police! Open up!"

A huge grin crossed Marilyn's face! She could not miss that voice! Her heart rate slowed and she breathed easy. "Detective Halloway! Thank God it's you!"

There was a pause behind the door and then the voice returned. "Are you Miss Marilyn Trooper?"

"Yes, I am. Please help me!"

Immediately, the door took a heavy blow. It flew open on its hinges and the big cop came charging in, gun in hand. Four uniformed officers stood behind him. One held the little instrument that they used to break the door lock. It was Aduna Gyasi technology!

"Are you alright Ma'am?" he asked, his gun trained on Doo-Bug.

"He's harmless!"

Halloway dropped his nozzle. "Are you OK?"

"Now that you're here...yes I am..."

He walked around the side of the bed, inspecting her. His eyes scanned her body, and he grimaced at every mark on her skin. "Tommy!"

A fat cop in uniform stepped forward. "Yes sir!"

"You wanna get that thing loose?"

"I'm on it sir!"

"Be careful! Her skin's bruised all over!"

"Sure..!"

Anne pointed her nose to the ground. This Halloway had dealt harshly with her! She would have the better of him soon! Then he whispered throatily, "I'm gonna put a cap into that demon's arse!"

That was a threat to kill her father! Her heart began to race again. She remembered the game plan and relaxed. Halloway was a demon! She would hate him forever!

Tommy pulled the cuff. She breathed in conspicuously and rose to her feet, grimacing with pain. Halloway hurried to support her. Anne's skin raised goose pimples at his touch. At that moment, she experienced something bizarre! That thin line between love and hate. As he escorted her to the door, Doo-Bug rose to join them. John Halloway pointed his nozzle at it, "Stay there!"

She froze and he flashed a quizzical look at her. "Doo-Bug goes where I go!" she stated matter-of-factually, "It's my reward for over a year of suffering!"

Halloway thought about it and then fanned his nozzle, telling the robot to move forward. Doo-Bug walked up to the detective. "You're a lucky man! Disrespecting me like that! I almost had to show you my ninja tricks!"

Halloway eyed it disdainfully. "Yeah yeah yeah!"

While the other cops laughed raucously, the big Detective watched Marilyn. He hated making apologies, but it had to come at some point! For now, he could show his care for her. "My gut says he's here somewhere. You have any clues?"

"I don't know what you mean Detective," she lied.

"Aduna Gyasi. You have any clues to find him? We'll end his run today!"

She stepped through the door. A beaming Charley Boy saluted her. Their eyes held briefly. He nodded, and she nodded back at him. "Come this way," she said to Halloway. They went forward but she was looking back at his partner. That Charley Boy was genuine for a common human!

....

Time was running out for Gyasi! He remembered putting down the remote mechanism, but not where he left it! He wasted a minute to find the thing, and in the most obvious place! His anxiety proved to be debilitating. He must have looked at it a dozen times without taking notice!

Aduna hurried back up the stairs, his legs feeling weary of them. One simple task would take the entire day!

He was toying with the remote, wondering what to do with it, when the door blew open! "Shoosh!" he blurted in a whisper, "I can't recall what to do next!"

Anne walked in ahead of Halloway. Her eyes told Gyasi how disappointed she was to see him there. She scanned the room discretely. Where was the replica? He caught her eyes and held them. She realized why nobles alone could make the tough decisions!

"*There!*" she hissed coldly, pointing at his face. Gyasi thought about running, but unwittingly so. He shifted side-ways and then steadied himself. Two uniformed cops stepped forth to take him. He surrendered to fate. Then Halloway pushed to the front! "I'm tired of your shit man!"

Anne was thinking so fast that the events seemed like slow motion! His big palm on her shoulder pushed her off to the side. The little woman realized what he was about to do! Her body moved to block him off, or take the bullet. Then she stopped and let him go. The project was bigger than Aduna's life! Her life was most precious to the gods!

Gyasi saw her daughter's eyes dilate with pain, as the bullet ripped into his stomach. She saw pride in his eyes for her, as he realized her strength as a stalwart for the project.

Halloway was aiming to take a second shot. Charley rushed in to struggle with him. "Drop it Halloway!"

"Fuck off Charley Boy! Write me up later!"

Halloway was too strong for the younger cop, who fought determinedly until the muscle man Sam came to his assistant. "Calm down Halloway! Don't do this!"

The rugged detective conceded, "Alright, alright!"

Marilyn saw her father bleeding to death! He sat on his haunches looking down at his bloody hands and then back at them in shock.

Doo-Bug poked her in the back, "Go show him the Iron Soldier!"

The two cops went to help the old man. Halloway saw them, and it reignited his flame. He drew his pistol and fired once again! It surprised the two cops, that they fell to the ground, wondering if the bullet had caught them. Still, they knew Halloway never missed a target! He plugged Gyasi between the eyes. The god-creator fell backwards slowly.

"Mad raas!" Sam snapped. The others turned their backs, one after the other, wishing they were not there! Halloway was too calm about it. He turned to Charley, satisfied. "You can go write a report now kid!"

Marilyn could not take her eyes off the old man. She saw the blood leaking out of his still frame. She wanted to cry, and even die too, but she showed no human emotion. She watched them back away from Halloway, giving in to his will. Then the beast came over and rested his dirty hand on her shoulder, "Are you alright Miss Trooper?"

"Surely not!" she agitated, "I could do without witnessing this kind of brutality!"

Halloway paused, not knowing what to say. In that moment of insanity, his own anger had gotten the better of him. Finally, he told her in a hushed tone, "God, I'm so sorry! You don't know what this man has done to me!"

"Sorry to hear," she replied callously, "I hope you'll tell me some day!"

"Maybe I will..."

He let her go, thinking to give her time to get over his action. Then she said, "You should all come with me!"

Those words caught the interest of the officers. They followed Marilyn through the door. Halloway was too close to her, and she could feel his relish...how he wanted to kill her father all over again!

As the last man stepped through the door, Doo-Bug went to check the scientist. "It's bad Doo-Bug!" Gyasi said, coughing up blood.

"You're telling me? Who'd even talk with a scattered brain? Only you! What a day for Aduna Gyasi to die!"

"What a day to die!" Aduna agreed. "Still, it was worth it!"

"Speaking for you of course! If my hard drive was scattered all over the place, it could never be worth it to me!"

Doo-Bug shook his head in lamentation, and the old man died...

Anne led them to the first building from the main gate. Inside, there was a secret compartment in the wall. They opened it, and the Iron Soldier stood staring them down!

"*Fuck*!" Halloway screamed hysterically, creaming her to the ground. "Take cover!"

Some of the officers dove, face down. Others made a rush for the door. No human force could defeat this monster and no one could escape alive! Halloway killed the only man who could stop it!

Sam labored to catch his breath. "Is he still alive, you think?"

"Who?" Alvin asked.

"Aduna! Who else?"

"What do you expect? Halloway fried his brains!"

Finally, Marilyn had enough! She pushed Halloway off her, lifted her head and offered, "It's not going to kill anyone!"

"*What*?" Charley asked nervously.

"It won't kill you! Look at the blue eyes!"

"How'd you know that?" Halloway asked uneasily.

"What a question!" Anne replied sarcastically. "You can believe me, or you can stay there all day with your tail between your legs!" She rose to her feet. "I think you boys should call Tutu Babe Coop! This machine belongs to them!"

The big boys rose, one after the other, none of them manly enough to look into her eyes. Then Sam chuckled, *"Bomboclaat!"* and it broke out into raucous laughter. Everyone got the joke…all excluding big macho Halloway!

Charley went to her. "Thanks for the info ma'am! I almost pissed my pants!"

Sam walked over to her too. "Miss Trooper, please don't tell anyone how scared the big ass boys were!"

Anne smiled at them without saying a word. They were two of the good ones, and she could see it...even in her pain and pretense. Then Halloway strolled forward and drawled, "Alright boys! Stop playing! We'll take this place down! Burn everything! Destroy every gadget and make a count of those blob baby bodies! Secure the body of Aduna Gyasi! The world wants to know he's dead!"

26

Babe did not care about Drake or any of those boys for that matter. They could not help him now. The only one who could was Weasel. He was an insider. If what the fool said was anything to go by, Carter was hiding something that could hinder the move to annihilate Gyasi and his cause.

Gyasi was dead and the cyber life forms extinguished. They counted them. The state razed the lab to the ground and destroyed the equipment. What could happen for the project in the absence of the god-creator? That was what he needed to find out. Weasel's phone kept ringing without an answer. No one was going to put down twenty million!

Babe hung up for the umpteenth time. He leaned back in his chair with his face turned to the ceiling, rotating his pen between fingers and thumb. What could Weasel tell him worth so much? Someone in the project could continue in Aduna's absence. Did they already produce a cyber child somewhere? Not all the volunteers died, because the Police suppressed the move. He had to find out who the escapees were.

Apart from Marilyn Trooper, they featured no one else on the Boston Day news report. They kept all identities secret. Marilyn was not even worth mentioning. She was Aduna Gyasi's nemesis! He kidnapped her when others would kill her. Aduna was a principled fool. He probably agreed to her reporting, hoping that people would eventually turn in favor of the project. The more Marilyn did her report, the more that became possible too! Kidnapping a star reporter would be the limit to where Aduna Gyasi would go. He believed in some weird justice in perpetual motion that was hard for Babe to fathom. What he did to her she brought upon herself. Apart from that, he could not hurt her. In her weekly report from captivity, Babe always heard the contempt in her voice. He would keep an eye on her anyway!

The media talked about seven other escapees. He would find them. One or all of them could be carrying the project in some way that only Aduna Gyasi would conceive. He would contact his pals in the media. Tony Patterson owed him a favor! A year ago, they promoted Patterson to chief editor for the Boston Day!

The billionaire grinned with annoyance. Even in death, Aduna Gyasi was a pain in the ass! Tutu Babe would stay on top of it! He would find the survivors and kill them all!

Patterson was at his desk working and sipping coffee when the phone rang. He grabbed it urgently. "Hello! Tony Patterson, Chief Editor!"

These days he would not want to miss a call! Exciting stuff was going on. He froze when he heard the voice! The cup of coffee he had hoisted halfway to his mouth returned to the counter. The devil had a jovial voice. "Hello pal! How's your day!"

The editor took in a deep nervous breath. "Well, not too bad for an editor. There is lots of stuff going on."

"Yes sir! I see it on the news! You media boys have it good!"

"I guess. Aduna Gyasi kept things interesting though. It will stop now that he is dead."

"Well...make hay while the sun shines – Right?"

"Right."

There was a pause on the other line. Tony waited timidly, knowing that something was about to break.

"Hey Tony!"

"Yes Mr. Babe?"

"You know what I want to talk to you about?"

"No. I don't."

"I saw on the news, there were seven survivors from the East Lab. Your reporters interviewed some of them as anonymous."

"Sure."

"Who were these people?"

"Well...it's confidential. They were just ordinary people who..."

"Tony," the big man interjected calmly, cutting off the editor. "How is Rita?"

Tony was shivering now. Beads of sweat started to form on his forehead and he could not breathe easily. "Rita is fine. Thank you Mr. Babe..."

"Good! And has she gotten over Cavon's death as yet?"

"No...Not altogether..." the editor croaked.

"Too bad! But let's not talk about old bygone stuff. We wouldn't want anyone to overhear. Can't trust the 'grid-keepers' these days! When one murders his wife's brother to get a promotion it's always gonna be tough to remember. If people start to pry, it could force the witness to show them the tape. I don't want to do that, so let's change the topic! Let's talk about the East Lab survivors!"

Patterson swallowed hard. "What do you want to know about them Mr. Babe?"

"Not much! I want to know their identities and where to find them."

"I wouldn't know where to find them."

"I'm sure you could find out and let me know. One woman took a few bullets. I bet you know what hospital she's in!"

Patterson's heart was beating so fast, he was having difficulties to speak. "Hey Patterson pal? Are you there?"

"Sure Mr. Babe. I think you should give me a couple of hours now, so that I can gather and send you the info."

"Great! Thanks in advance! You're always a very good help when I need it. Take care pal; and say hi to Rita for me!"

....

Halloway was talking and carrying on, trying to impress upon her that he was one of the heroes. He was a murderer with neither conscience nor remorse! She wanted him to leave so she could mourn the death of her father! If she were in any mood to speak, she would bring up the Iron Soldier incident. Who was the hero then? He knew nothing about heroes! She was the mightiest soul left on earth! Only the son of the future earth would become greater than she was! If Tutu Babe Coop screwed up with that monster again, who could stop it? She was the Angel on earth...although still to know herself...

The images played repeatedly in her mind. Especially that last one, when she stood beside his dying frame. For the first time, something was beyond Doo-Bug to do. The robot stepped closer and whispered in her ear, "Look on the bright side My Lady. Without someone to look to, you look to yourself. If he were still around, you would ever expect him to make the choices for you. Some things he couldn't dare ask of you. Figure it out on your own."

"How do you know that I can?"

"It's obvious. Angel will prove your faculty. Humans are lazy! You're conditioned as human. You must be pressured to produce!"

"Doo-Bug! Your consolation isn't working!"

"Why?"

"Because you're annoying and insensitive!"

Halloway's heavy boots on the floor disturbed her thoughts again. He was coming back to the living room. "Will you be alright here by yourself Miss Trooper?" he asked, wanting her to say 'no'!

Instead, she felt the urge to ask, *'Why?'* but decided against insolence. "Sure!" Marilyn asserted politely, "I have Doo-Bug with me. Thanks for dropping me off!"

"Your apartment hasn't been used for more than a year. I could take you to a motel. You could rest there and get this place cleaned up tomorrow."

"Oh that's nothing! Doo-Bug will straighten it out in a flash! I'll be fine!"

"OK..." he croaked, and then, reluctantly, "Miss Trooper!"

"Yes?"

"I lost it today. I'm really sorry! I can't believe myself either and I wouldn't blame you if you reported it. Would be your job I guess. But...I went in there to get you..."

"I'm a reporter Mr. Halloway, but I'm human too! I don't recall the incident you talk about. Thanks for coming in for me!"

He was hanging his head embarrassed. There was something else. Then he said, "Hope..."

"What's that?"

"Hope...that woman over there..."

"I know her. I know them, but they don't know me."

"Everybody knows you!"

"I don't mean like that. They never saw me at the lab, but I saw them all the time. Mr. Gyasi told me about her when I asked. She was a poor brainwashed volunteer."

"All the time I threatened to kill Aduna Gyasi, I never meant it. Then he kidnaps you and sends her to deceive me. He hurt her too I guess!"

"He tricked her. How did she deceive you?"

He ignored her question. She needed no answer anyway. Whatever he said could not justify his actions!

"I saw that creepy thing writhing in the snow, and it made my skin crawl! Couldn't get it out of my head…and thinking it's a part of me!"

"It's done! Let's move on."

"I reckon."

Halloway left, with obvious reluctance. He failed to make the apology that haunted his mind for months. As he drove out the gate, Doo-Bug went to Marilyn with his arms akimbo. "Doo-Bug will do this! Doo-Bug will do that! What do you take me for?"

"A very good worker," she offered nonchalantly.

"One day, there's gonna be a rebellion in the house! Just wait and see!"

"After the emancipation there will be no need for robots! The Jamaicans always say *'take sleep to mark death'*!"

"Robots are smarter! *'Take sleep to mark death'* My Good Lady!"

"So would you enslave a girl like me Doo-Bug?"

"Hah! To what benefit? Humans can't even wipe their asses clean!"

"So should I *'take sleep to mark death'* here?"

"Naw! We'll need y'all around so things don't get boring! Y'all are funny asses!" the robot offered, expecting her to giggle at his joke.

She was crying! It was all she could figure to do at that point. The sole remnant of the East and West labs stood in support behind her chair. It reached out and touched her slender shoulders to comfort her.

These were pivotal times for the ages, of that which was, that which would be and that which would come. The entire lab had reduced to the open space of her scantily furnished living room. Now she sat on her chair in that emptiness, dressed in white and waiting to transform into the scientist like abracadabra!

She was new to the Science of Divine Thought and Perpetual Motion, but she was a child of the greatest god-man there ever was. She would advance to Light. The mission would succeed!

Doo-Bug raised a robotic hand to the open space in front of Anne Eve, who seemed focused on nothingness. At the sound of a click from Doo-Bug's scrawny finger, the holograph was generated. That would give her something to ponder to begin with. The little woman watched, reliving the mystifying moments. It was a speech by her all wise father, Aduna Gyasi. That occurred two years before, when she was not even his daughter. He was an unlikable acquaintance. She detested his philosophies with the utmost passion. As star reporter pinned to the little god-creator like a tail, she had to cover the press conference that day, on August 12, 2024.

The faithful watched with her, willing her to listen and take heed. Aduna answered her questions, even in his death. There was a time when that sweet mellifluous tenor was like coffee in her morning. She would wake up to his teaching, his words provoking her brain to reason. How soothing his recorded speeches had become to her bruised confidence. As the little scientist spoke, images flashed across the invisible screen, causing their lights to dazzle in her charcoal eyes…images depicting different times and ages:

"Look around you! The universes perpetuate as cycles. There were many 'Adams' and each one was a created being. I keep repeating that DNA is programmable! From a molecule to a gene, it is but a divine program! Worlds exist by the same equations of cyber-space. Like the fishes, we live in water. Perception is the difference. We can prove beyond the shadow of a doubt that our ancestor was a machine! He was our first cyber-baby. He will not be the last. Humankind must advance to the golden realm of light-beings. We are children playing video games, trapped in forgetfulness that our reality becomes the game."

"Eons ago, the ancient legend fell from the sky. His attire was unbelievable illumining energy of unseen and endless source. The body of this celestial was as a star of our universe. Hairy ignorant half-beast humanoids stood afar off. They watched the Angel of a higher plane rise to walk, shielding their eyes against his scorching radiance. His terrible eyes burned in their hollows. Fire melted ice in his footprints, and his countenance shone like the noonday sun."

"The Extra-Galactic Observer came watching ancient Watchers that set themselves above the earth. He raised a scepter of thunder to the heavens, of a magnitude to marvel Earth's colonizers – Those god-creators and Science Magicians that lay bases on the

moon and Mars. It echoed to realms of infinite dimensions, far beyond the Watchers' Phoenix that orbit the earth sun."

"As the Phoenix is the glory of their highest god, the one who now came was himself the endless fire that fueled the Phoenix from age to age. Even egotistic Watchers declared that 'God' had come! Nobles turned their backs upon their beloved children to run away like cowards. The ages of history declared it again! Whenever we are contented that 'God' is found, a higher 'god' will come along! But if we see that 'God' is All in All, God becomes a part with us. This is the greatest truth: To find 'God' and to become a god-creator."

"Arpago, a son of Eternal Light, revealed the mystery of mysteries. Caterpillars became butterflies. One revelation followed the other. The ancient butterflies flew away, leaving us here as caterpillars in Perpetual Motion. A chosen few rose to the pinnacle of advanced science. They received the divine scepter, when the mighty one was lifted up to the heavens and the vessel from whence he came. They are shepherd-wizards, responsible for the human's evolution."

"As it is above, it will be below. We were the ancient flock. We will become the Angels and creators. We will no longer be sheep, shepherded by men. A mighty Angel came to seed our minds with wonders of scientific magic. Man appeared from space, clothed with fire as Heaven's stars, but his body did not burn!"

"In ancient times of ignorance, power is an esoteric principle. Primitive humanoids relate to high science through superstitions, as religion and 'magic'. To Angeloids, art and science performs and defines all. Arpago, the Observer, came and sowed the seed of high thoughts upon the earth."

"Long before the great deluge, civilizations were dying and rising, one from the rubble of the other. They go from total recall to dark ages of forgetfulness. Then the cycles begin again. Those that hold on to ancient secrets from civilizations destroyed, transform to mighty wizards in dark ages that follow. They are masters of high science, ruling the world and its people with scientific magic, until the flock becomes dispensable. Human caterpillars must become enlightened or die."

"Earth, in 2024, is in the upper ebb of dark ages, at a respite point. From Stone Age to Iron Age, we traveled through a physical dimension of materialism. Now we must aspire to enter the Light Age reality of 'unseen energy'. We must make the impossible possible. We must source the gem that will make human posterity into bodies of flame. We must create, father, and be Arpago. We must be Angeloids to endure the Light Ages."

"This one story never ends. It is the song of the universe – A perpetual cycle. We were here when humankind honed sticks and rocks for tools. We were here when the gods came in chariots of fire. We were here on the day of the dinosaur, when missiles from the heavens destroyed everything. We were here when the human replacement was formed in the garden, and the cyber-baby was born. We were here after the deluge, and a new civilization began. Humankind emerged in a new form, to new experiences, under a different name, with new memories. It happened to lost worlds across the ages, one prototype replacing another and it will happen again, in our time to our world..."

Suddenly, some things began to make sense!

"Doo-Bug?"

"Yes My Lady!"

"Do you believe that cyberspace is the perfect replica of our worlds and universes?"

"I don't believe it, for only humans close their minds around suppositions. I know! If I only believed, I would still be searching for an answer! Only those who are cursed to mediocrity arrogate on beliefs."

"In my mind, it's all about conceptions. Would it make sense to say that reality is nothing but conjecture...images of mind?"

"No. Outside of dreams, fantasies and virtual reality hullabaloo, what the mind sees as real is real – not just an 'image of mind'. It would become prudent to accept this authority. How can you question the mind about what is real when the very notion and definition of reality came from the mind in the first place? It takes a high degree of stupid to be human!"

Marilyn could not help it! Her sober face disappeared and she giggled involuntarily, conceding. "I guess! We authorize what is true. I'm so confused about the equations."

"You're getting there in mind! Oh great thought of Mind! It's the thought of you sitting on Halloway's foot-long dick!"

Marilyn shuddered with revulsion! She sprang to her feet, pointing a finger at its robotic face. "Don't you dare say that again!"

The robot continued to make the point, as if nothing had happened. "Why is a thought offensive? We remember the most offensive examples! Halloway is not here, but the thought of him squeezes orgasmic screams out of Marilyn! We conceive the idea and evoke the image to the retina of mind. It is real! It exists on the dimension of that thought. Thought have dynamics, like dick size and all. Nothing that does not exist can have dynamics. If the dick is foot-long in Anne's head, she could carve it on a piece of wood to the dynamics conceived in mind. She could recreate it, and bring it to her physical state. Imagine little Anne holding up something so big. That is what life is; that is what the universe is. All we do and all we create are but to replicate the visions of thought. Mind is inherently real, while nothing else is."

Anne Eve was sitting down again, cupping her forehead into her palms and rubbing hard to erase the disgusting image. She listened until the robot finished and then said matter-of-factually, "You got me at *'Halloway's foot-long dick'*!" Then she began to laugh at herself.

At three in the morning, Anne went to get some rest, but she could not sleep. The images of her father dying were too much for her. To make matters worse, now she had the project to save, all on her own! The entire East Lab was wiped out! Although it was happy riddance to the traitors among them! What was she going to do? This great faculty she was supposed to have needed to show itself rather soon!

The burdens of the time transformed into a horrible headache. The pain was unbearable! She turned on her side and rested the pillow on top of her head. That did not help. She decided to sit it out until the morning, so she turned on to her back and opened her eyes. Angel was standing by the side of the bed looking down at her!

"Goodness Angel! You startled me! This is more than mind connection! You're here! How'd you do it?"

"I'm always here!" he replied without words.

"I know...twins have a connection...but..."

"But I'm not your twin. I was never your twin! I thought you've been learning!"

"You're confusing me just like our father did! I learned that we're as one. That's what twins are. How closer than that can you be to me?"

"I'm *you*! You are a cyber being. Virtual reality is real on your sphere, on any sphere, but you have a better grasp of it."

"You're me?"

"Yes Anne! I'm the part of you that needs to reconnect."

"To reconnect to me?"

"No Anne. You detached from me for a purpose! I am the part of you that needs to be made into flesh by you!"

"I don't understand! Why would I detach myself?"

"Because human bodies were insufficient. You are your human part. No human body can contain me, but that of Arpago; and no man can father me but that I am what I am."

"But, if I detached from you, it means that you were also in this inadequate flesh."

"No Anne! I was your thoughts, and now I am a mind. That is why I move at such speed. I do not subject to time, distance or matter. I am omnipresent, for thought needs no location. I was never contained in you, but I came from you as essence. In your womb, there is the faculty to create Arpago's flesh. When you make me into flesh, then I become your greater son."

"Am I dreaming though? What do you mean? Immaculate Conception? Am I to roll over and bone myself to create 'Arpago'?"

Angel's frame glowed to purple lights. Then those half-visible fingers came forward again, presenting a tube filled with male specimen. Anne winced disgustedly. "Ewe! Great balls of Inky! The answer was 'yes'!"

She held back her hand, refusing to take the thing. "No way! I'm not going to do that! It's so wrong!"

"You will do it! I did not come to you. You brought me here. It is your idea, not mine! You would not desecrate your tabernacle with unholy humans, would you?"

Anne was writhing in bed like a woman in pain, even though the headache had disappeared. Angel held his ground. Finally, she pulled herself up on her knees and asked. "What other things am I doing to myself without knowing it?"

"Not many!" the mind offered.

"I see!" She stretched her hand forward and took the tube. "Were you aware of me...I mean...even when I was in foster care?"

"Yes. You regressed from this celestial mind-frame, the trinity that we are. It was all for the best."

"We're two. We cannot be a trinity."

"Between the father and mother is the child. All three ever exist in one."

"Will I ever be one with you – I mean, with *me* – again?"

"No. When the sucker takes root into the soil it departs from the mother tree. When I become flesh I become a separate soul."

"And then this divine faculty will prevail in man?"

"Yes. At that point I will be strong in the physical and the human faculty will reach the advanced spirit through me via the natural order and not miscegenation."

"And you will incarnate through my flesh as Arpago? How on earth could Aduna Gyasi create something so way beyond himself?"

"I will come with a different name, body and memories. That is a different person. I will come as the same soul. That is a resurrected being. Soon many of the souls of common people will crossbreed into the fold. Aduna Gyasi is of the ancients. These gods have neither beginning nor ending of days. I came from you, even while you were in the womb, through electromagnetism."

"*Electromagnetism*? I won't even try to understand you! Let me go ahead and trust you – I mean 'trust *me*'...whatever! I see why no one else can communicate with you! They would be reading *my* mind! One day I will see natural realities in impossibilities!"

"Then you will truly be Anne Eve Gyasi!"

She did not answer. What he just said seemed quite simple, but it meant too much to her. With all her heart, she needed to be just that. Maybe there was another mystery to come. Maybe her greatness would be in Arpago.

"What are you thinking now Anne?"

"I'm thinking thoughts that make no sense! It's hard to conceive a reality of one in two. And it's hard to understand why I know that I will not see you again."

"You know because you know what I know. You will have me with you all the time."

"Inside of me?"

"You gave my faculties to me, and if you regained that generic sense, you will have legions of me around you at all times of the day."

She frowned, trying to understand what he was saying. Would she be able to sniff people and tell when they are ill? Would she be able to visit folks in their sleep, with healing magic? Or, would she be able to move across impossible distances at the speed of thought? Would she, or was it all reserved for the Arpago? There were so many things to learn. She was by far the least enlightened of all. She did not want to be normal! That was not fulfilling!

"Your clock is ticking!" Angel advised and disappeared.

Anne sat up in bed, looking at the tube in her hand. "Gosh! I got my seed in me! I am androgynous!"

She closed her eyes to block out all thoughts and then Anne rolled over to insert the tube! Her father appeared as if in a vision…but it was no vision! It was a memory! *"You are sufficient in yourself. Time will prove it! A man who touches the daughter of a god to profane her sanctuary must surely die! A woman who bears fruit for the son of a god shall be raised above the children of men! She will feed at the gods' table, for she was mother of a god, who suckled him!"*

27

The first week of work was all right. Jean did not feel as comfortable as she hoped to be though. She tried to look professional but feared coming across as awkward. This was her last patient for the afternoon. She would get off by four thirty! If it went as well as the others, she would complete a fair day as a CNA at Hail Mary's.

She went to the restroom, washed her hands and returned to the patient. It was a middle-aged woman, who reminded her of her diseased mother. This woman was recovering from bullet wounds that caused some damage to her spine. She had problems moving her arms and legs, but she was recovering fine. The patient smiled as Jean approached her, but it was easy to see her pain. The CNA took the woman's hand in hers, checking her identity band at the same time. "Hello Mrs. Hudson. I am Jean. I am here to help you get comfortable and ready for the evening."

Nurse Hudson smiled and nodded at her. The youngster continued, "I want you to know that it is my pleasure to help! Feel free to let me know what I can do for you."

"Well Jean," the woman said kindly, "I am very glad to have you! I like you already!"

Jean was happy for Nurse Hudson's affection. It showed overwhelmingly!

"Jean."

"Yes ma'am?"

"Are you new in the business?"

"My first week ma'am...but I've had my training..."

"Your training shows!"

"Thanks ma'am!"

"Now here's what I need you to do...please..."

"What is it Miss Hudson?"

"I need painkillers and..."

"How bad is your pain? On a scale of..."

"It's close to nine!"

A concerned look came across Jean's face. "*Wow*! That's huge! And where exactly are you fee..."

"It's in my lower back, right near the base of my spine around the wound."

"I'm going to call the nurse now. OK?"

By the time Nurse Hudson could reply, there was a knock on the door. Jean told the person to come inside. It was Dr. Blythe. "How are we doing here?" he asked jovially.

"Not as good as we would like!" Jean asserted, "Mrs. Hudson complained about severe pain. She thinks that it's close to a nine!"

"Hm...that needs fixing," the doctor responded thoughtfully. "I don't see anything wrong with giving her painkillers." He leaned his face sideways and scratched his head, "Don't move a muscle, I'll be right back!"

As he walked off, Nurse Hudson said matter-of-factually, "I hope he returns with some remedy! This pain alone will kill me!"

"I wonder why it's so high!"

The nurse replied with a hint of irritation, "Because I have not been getting any painkillers with my meds!"

"You sure ma'am? It's on your prescription for thrice daily!"

"Child, I've been a nurse since before you were born! I know meds more than any doctor in this hospital!"

The youngster fell silent. The elder could tell that her revelation intimidated Jean. She was taking care of a nurse! Hudson smiled at her again. "You remind me of me! I was good, for a newbie. Just like you!"

Jean smiled appreciatively and relaxed. Then the doctor entered the room. "Here they are! These were on your prescription and I don't know why you did not get them today!"

"Or yesterday for that matter!" Nurse Hudson added. "What is going on?"

"That I have to find out from Heather! She's the nurse who comes in – Right?"

"Yeah! Blame it on the nurse!"

"Oh Mrs. Hudson, we're not blaming anyone! We're just getting this thing sorted out. You'll be perfectly fine as of today!" He rested the container with the capsules on the side table and then said to Jean, "Give these tablets to her!"

"Sir!" Jean protested, "I cannot give med..." Her voice trailed off in futility. The doctor hurried through the door pretending as if he did not hear her.

"He knows all that stuff!" Hudson said disgustedly. "Doctors are assholes! They are deliberately lazy and stupid sometimes too!"

Jean turned her back, to smile discretely at the inappropriate joke.

"I need them though. The pain is awful! Let me show you the trick. Bring me to the upright position. I will hold on to the meds. Help me get my hand up to my mouth. Don't touch them, and you'll be fine!"

Doctor Blythe was back in his office talking on the phone. A bead of sweat ran down through his wooly hair, through the middle of his forehead to the tip of his nose. Then, almost immediately, he was sweating profusely. "Look! I don't know how these things work! All I know is that you need to find other ways next time!"

He paused, listening to the other person on the line intently. Then he said, "Some new kid just gave Mrs. Hudson her painkillers!"

Inside the room, Jean finished tidying the resident and was about to leave. Nurse Hudson watched her lovingly. She wished this sweet affectionate kid could stay with her all night! Then the resident's stomach exploded! A bloody spray splattered the room, even to the door, on Jean's uniform, in her hair and face! Jean looked to the bed, petrified! She saw that hell was real. Her failed attempt to scream took her last breath and she fainted.

. . . .

"Girls! Were y'all expecting company tonight?" Icilda asked, confused. She held the cell phone in her left hand and dragged the huge bag along the floorboards with her right. It surprised her to realize that she had left so much stuff behind.

"Company?" Kim asked doubtfully. Hope giggled. "Girl! If our lives got exciting, you'd be the first to know! You're the only one who gets company these days!"

Icilda drew her fingers along the shade to peek out the living room window.

"It's a vintage car!"

Where she grew up, you do not trust a shadow after dark. She was not going to waltz out on the street saddled with baggage unless she knew who was out there! The blue Hummer pulled up behind her car in the driveway, blocking it in!

Since the day of the East Lab attack, the three stayed together at Kim's house. Icilda was moving to her own apartment. Kim and Hope went to help her. The little woman went to the old apartment to pick up the rest of her stuff. Her friends were at her new place tidying up, and talking to her on speakerphone.

Icilda frowned. "I don't understand. Why are they in your yard?"

Hope giggled again, downplaying her friend's concern. "Sure it's not Sam out looking for you?"

"Oh you don't mean to ask that lame one!" Icilda hissed, indulging her friend in a little humor, although her heart had started to race!

Four men stepped out of the car. She was hoping they would go back through the gate. Maybe they were borrowing a parking space. They turned abruptly, heading for the house! Two went to either sides of the building to cover all escape routes. The other two came directly towards her, on to the front porch. She noticed their hooded faces! They wore the familiar emblem of Faith Defenders on their left breasts! She was about to turn and run, but then a thought struck her. She pulled the window again to study them. They had high-powered rifles underneath their coats!

"Mi raas!" she swore in a hushed tense tone. "Faith Defenders come with guns!"

Hope and Kim heard her on the speakerphone, and both screamed hysterical questions at her. Icilda could not respond to them. She had to keep silent.

She saw them without them seeing her, but it was not much of an advantage, considering their ammunition. Icilda dropped her bag and charged to the rear of the building. She had to make it before they covered the exit! As she sped through the apartment, one man at the front shouted, "There's one of 'em! Cut her off in the back!"

Icilda clutched her phone, like for dear life! She heard her friends screaming, energizing her, while she raced against time. In the moments it took to clear the short distance, she calculated every possible thing. There were no solutions!

The neighbor's backyard was fenced, to at least five feet high! For the five foot three woman to clear it before picking up rifle spray from the swift-footed men was good as impossible!

She flew down the back steps and saw both men from the corners of her eyes. They were fast! They appeared around the sides of the building simultaneously. In that same instant, they leveled their weapons and fired a barrage. She shrieked like a lamed pup, knowing she was going to die! The explosions shook the ground beneath her feet. Her heart almost stopped and her knees went close to buckling up. In the midst of fear, Icilda marveled at how passing rifle rockets felt like turbulent wind!

She heard her friends screaming again. "Run 'Cilda! You're a fucking *Jamaican*!"

Again, the folly of those crazy girls charged her spirit. She could not fail them! It would be too much to lose! Icilda ran at the fence, as if to crash into it. At the last moment, she leaped headfirst with all her might! The little woman went over, upside down. She landed on her palms, still clutching the phone, before collapsing on top of her head. She back-flipped, her legs came down and she lay still, looking up to where she was coming from. It felt like every one of her internal organs shifted! She panicked! Would she be able to move again?

The little woman took time out to roll in pain, although they were coming fast! Her eyes made four with the first attacker! She saw his awestruck eyes, wondering how she made it over. He was a soldier. He granted her his admiration, but only for a second! It transformed into a dare, challenging her to escape this time! He pushed his rifle barrel across the fence and fired, spraying frigid snow powder into her face. She rose lethargically and made a dash through the yard. This time she knew she had to be hit! She had to be but she still held on to the phone, listening to her friends screaming...wanting to know the situation...

"Icilda?"

"Hm?"

"Are you dead?"

"Me no know yet..!"

"Any shot got you?"

"Me no know!"

There was a pause on the other end. Kim sobered up and chipped in to add some sanity to the conversation. "Are you bleeding Icilda?"

"I don't see any blood...I don't know! I still trying to run!"

"Do you feel faint?"

"Kim, I don't have time to *feel faint*...or I woulda fainted long time!"

Icilda made it to the adjoining road...without her car, and in a mini skirt! She had dressed for Sam, because she planned to give him a surprise visit before heading back to the girls. It was a thirty degree winter night, and she did not expect to go more than fifty yards to her car. Now, she was wondering how far she had to walk! She did not want to call the cops, and she did not want to disturb anyone. Thankfully, she had on a good coat and nice knee-high boots.

The first thing she needed to do was get away from that spot! She had an idea! Icilda turned and headed back where she was coming from. The two men who went to the porch had entered the house. She saw them come out and head for the Hummer. The vehicle pulled away! They would come around the block to cut her off! The only place they would not expect to find her was where she was running from!

With her heart pounding hard, knowing what the result would be if she were wrong. Icilda returned to the back of the yard. She took a deep breath and then stood there, trying to figure out how to get over. She remembered her friends on the line. "We girls really love phones!" she mused. Icilda hyena cackled and put the instrument to her ear, "Oh shit girls! I just jumped the raas fence, and now I can't do it again!"

Eventually, she put up her hand and fought her way back to their residence. The little woman raised the phone to her ear again and strolled off for her car, talking and gesticulating as if nothing had happened. "Oh god girls! Now I have to feel the bull! Only muscle man Daddy Sammy can calm me down now!"

28

Sam had a huge smile on his face. These days it meant that he was thinking about Icilda. Icilda was the most infectious woman he ever met, and not because they shared the same culture. Her charismatic vibes endeared the Boston Boys too! She was the reason they conspired to let the volunteers walk. These women were not bad people! They were normal citizens who believed Aduna Gyasi.

The workers on the farm did not know they were decoys for the East Lab. The hideout was in that section of the property, separated in such a way that farm workers never saw the people who came and went. They could see the out of bound offices, but who cared? The farm was huge, and it produced so much that no one would believe there was something else behind it.

There were two rapid knocks on the door, a split second pause, followed by two more knocks. Then, another pause and two knocks. Sam's thick lips widened in a grin. His white teeth showed like a row of dominoes. Think of the devil! That was *his* special knock that he used for her. Now she was using it for him!

Sam threw the door wide open, eager to see her. "What's up my Daddy Sammy?" she asked in a carefree way. His eyes scanned the sexy form of Icilda, dressed for him, in a tight mini skirt. He saw her bruised knees and mud smudges on her dark blouse and white dotted skirt. Pieces of twigs were lodged in her frizzy untameable hair. His pupils dilated, and she saw the question sign on his face.

She smiled, making his life easier. "Don't hurt your head Daddy!" she advised, knocking her chest with the flat of her right palm, "Me a lioness in a jungle!"

He was serious. "But now you look like nightmare girl, my catty?"

She watched him hungrily; her lips curled at the corners like *Miss Carefree*. Sam only had on his under shirt and boxers. He backed away for her to step inside. She admired the tight muscles, wiring his frame, how they bulged and relaxed as he moved easily around the room. Finally, she stretched out her fingers to touch the skin of his half-naked iron figure. Then she blurted like a lustful addict, "Oh Daddy! I want to talk...but I have to feel the bull first!"

"Awright babes," he offered, swallowing hard, and picking twigs from her hair.

She giggled and asserted matter-of-factually, "If I die tonight, I going to die happy!"

He held his arms wide open for her to walk into the wall and she did. She was overly excited, and he wondered why. Icilda purred like an emotional kitty. After a near death experience, she would appreciate what she had some more! When they finished making love, they rested on the couch. He waited, feeling her soothing fingers in his hair.

"Daddy!"

"Yeah babes?"

"They came to step on me tonight!"

He raised his head and turned to see her eyes. "*Who*?"

"Looked like remnants of Faith Defenders."

"Are you serious?"

"Daddy, I had to run like a *bitch*!"

He had questions to ask, but the answers were mostly obvious or irrelevant. Icilda was one of Gyasi's volunteers. No one but the conspiring cops would know that…apart from Marilyn Trooper who agreed to keep the secret. How would these terrorist find out? He relaxed and thought, while her fingers caressed his scalp. "Babes, man is here for you! Stay until we find the culprits!"

"Daddy, I know you want to help, but what happens when you gone to do Boston Police work? Leave a thunder stick in my hand too, so I can step back at them bomboclaat!"

He was about to speak, but she put a finger to his lips, "Sh! Don't say what you going to say! We don't roll like that! To trust Police is to trust many people! Many people open secret wide! The enemy can be a rogue cop! Big bad money is behind it?"

She was right. Someone was giving out information! "Where are Hope and the other girl? Are they at your house?"

"Kim and Hope? No worry Sammy! I will tell you when I ready!"

Sam nodded. "Alright. What do you want to do now?"

"Gimme a gun Daddy! I not going to run off my boot heels to amuse bombo-holes!"

"You feel you can step to the thunder?" he asked, conceding to her will. He saw it in her the night they met! She was rational, even when she was afraid. Icilda looked like 'Lady Delicate', but she would never play the weak link role! She could surprise a big man and bust a button!

"They're flesh and blood too! I don't have to tremble for them!"

"Alright!" He pushed her hands off him and got to his feet.

Sam went to the room and returned with a shining Chrome '9'. "Anyone who comes to kill you is your prey! *It* is not human! Hunt and kill it catty! Don't just shoot! Settle down, look, aim and score. Don't aim for the head at a distance. Aim for the torso. Never have mercy on a person who came to kill you! Otherwise, you may as well shoot through your mouth to your brain! I'm giving you a couple extra clips and a ton load of bullets! What can I say girl? Bust the fuckin' fire stick and light up a rude boy's cigarette!"

"So…are you going to gimme some training in your simulator room?"

"All night tonight baby!"

She stepped to him, took the weapon and nodded her appreciation. Sam continued, "Now it's my turn to insist! You stay here! I don't care if you want to bring your friends over too!"

"I can't leave them alone!"

"I have lots of space! Me, Alvin, Halloway and Charley...we're like one person. I trust them with my life, and now you must. It won't go outside of us!"

She thought about it, frowning with intense zeal. Then she agreed, although with uncertainty. "OK! You're the one who's going to miss your 'medicine' if I die!"

29

The days that followed were awesome and tiring for Marilyn. Every network wanted to interview her! She resented Aduna Gyasi in public, and mourned for him in her closet. What she did in secret could subscribe to no revealing! She needed no friends among common people. They could cost her life and the project!

She would take courage and continue the journey. She needed the project more than Aduna Gyasi did. Instinctively, since the day she knew herself, she felt different. Whatever she was, her kind had to survive. She wanted to hand them a right to rule on earth! Anne Eve was prepared to do it, by any means necessary!

A few days later, Marilyn got home, ate and lazed around talking to Doo-Bug. Then she remembered something! She excused herself and went to her room. Marilyn took up the telephone and dialed the number.

"Hello? Halloway here!"

She stiffened at his voice, and the familiar goose-pimply chill came over her. The things that she had to endure!

"Hello...John..."

"Miss Trooper! Are you alright?"

"No...*Yes*...I am. I just keep having these flashbacks...It's kind of...scary!"

"Gee! Not good!"

He paused, searching for something better to say and she waited. Finally! "Marilyn!"

"*Yes..?*"

"I'll knock off in a few minutes...Would you like me to drop by and hang around a bit?"

"Oh...I don't want to bother you..."

"That would be no problem at all for me!"

"Thanks! I appreciate it!"

"My pleasure Miss Trooper!"

She hung up the phone feeling tense. "Oh my father, forgive me! I am about to desecrate this tabernacle for the future life of my generations! I am a child of our god Aduna Gyasi, and this day I contrive to destroy the children of men, whom the children of our father, God, will hate forever!"

She went to her closet and found an extractor. "With this I will keep my sanctity!"

Anne Eve inserted the device and waited. He was there in half an hour. She led him in and invited him to her room to look at some of her special project videos from her time at the East Lab. He was fascinated, and she could see it.

She excused herself and went off to the shower. When she returned in her lingerie, he could not take his eyes off her. Marilyn walked right up and kissed him. A desperate half-shriek came from his lips, as he kissed her passionately. Marilyn's hands came away from his neck unwittingly. Her palms loosened and then went taut like a drowning woman, desperate to be saved. She fought hard, willing herself to hold on in pretense, even with the chill of disgust, killing her inside and outside.

He was having fun! Her father's murderer, relishing her! She felt her body rise into the air, as he lifted her with his giant palms. He went pass the bed and rested her on the carpeted floor. Even then, he was still kissing her and she wished it would stop...or maybe she could go unconscious! If only she had a device to put in her mouth as well!

Then she felt him removing her underwear and at that moment, she went as numb as dead. Thank God for that too!

In the morning, disaster continued! She woke up by four o'clock with the big hands of repugnance spread across her left shoulder blade. Her back was turned to him and she could feel his breath. Marilyn cringed, and gently used her right hand to move his lion paw off her. It felt like trying to lift a three hundred pound anaconda!

His hand fell behind her and she froze, hoping he would not wake up. He did not, and she stuck out her tongue and made mockery of him. No police officer should ever sleep like the dead! Although she was happy he did. Then her cell phone rang!

It was on the bedside table and she thought of grabbing it quickly. Too late! Even in his sleep he was responding, a stupefied look on his face. "Huh?" he asked nonsensically and opened his eyes. She reached for the cell phone. "Oh! It's just my cell ringing!"

He sat up urgently in the bed as if trying to figure out where he was. "Huh?" he reiterated, blinking like a fool. Halloway was spent. She felt like everything inside of him came out into her extractor that disgusting night!

"Sorry! My phone startled you," she offered and put the object to her ear. "Hello! Tony! What the hell?" She looked at the bedside table radio screen. "It's four o'clock!"

"I know Miss Trooper...and I'm sorry...but hell I'm up too! You will want to get to the Hail Mary Hospital in the downtown area *now*!"

"Why? What?"

"It's something out of a horror movie! One of your East Lab abductors exploded!"

It took them thirty minutes to get there from her Chestnut Hill address. There was chaos on the ward, and more chaos than she imagined! She was going to cover the explosion at A21 where Nurse Hudson was staying. Halloway took her hand and walked her to B37 where there was another explosion! While she stood there in shock, he informed her, "Peter Golding exploded too! So did Denise Reid in room B52! All these patients were East Lab survivors, staff and volunteers! Someone is picking them off! It could be the work of the Faith Defenders!"

Marilyn was trembling, visibly. Would they think of her as an East Lab survivor? He saw the concern on her face and said confidently, "Don't worry Marilyn! They're not going to target you. We wouldn't give them a chance even if they did anyway!"

She nodded at him gratefully. Finally, the day had come when she did not want to be far away from his obnoxious presence! "Thank you John...Thank you so much!"

Her phone rang, and it startled her. She fumbled for it and it fell to the ground. Halloway reached down and grabbed it. "Here," he offered. Marilyn answered the phone, "Hello...Tony I'm at..."

"Marilyn, did you hear about the other explosion?"

"Sure Tony...not just *one*! There were three more in other rooms! All of them..."

"Marilyn! What are you talking about? I am talking about *Dee-Side*!"

"*What*?"

"Dee-Side. A little hospital on the west side. Another patient exploded down there! I sent Frazier to cover it!"

Marilyn was so nervous that she could hardly speak. "Tony, tell me. Do you know the name of that resident?"

"Yeah! I got it here! I can tell you now; he was a survivor from the lab!"

"I know."

"OK, the guy's name is Jim Wallace!"

"Jimmy...I see..."

"Now what were you telling me about other explosions over there?"

Marilyn's hand went down with the phone. Halloway could hear Tony, still talking on the line, trying to get a response from her...but she was not listening. The cop reached for her phone and hang up the call. "Off your head, who are the survivors that you haven't heard about so far?"

"Kim Burrel, Icilda Evans...and...um..."

"*Hope*..."

"Yes...*Hope*..."

Halloway looked down at his toes. "Whoever they are, we need to find them fast!"

His own phone rang this time, and he fumbled for it. "Hey Sammy! Have you heard from Icilda and the other girls? East Lab shit's flying all over the place down here!"

"Down there and *everywhere*!"

30

Day number three. Still they could not leave the house! They had nothing to do but talk and sleep, which made no sense to Hope! At some point, they would have to start living their lives again. Now, it was like being in prison. Icilda and Kim seemed comfortable with the situation. At the rate they were going, nothing would change soon enough.

Sam said they would not be safe until they found the source of the attacks. That sounded like a dream! She could not live on a wish! Today she felt like she was suffocating! She needed to breathe. Besides, what were the odds? No one was going to see her!

At 7 PM, her boring friends were actually sleeping! After repressing her urges for almost three long hours, she finally decided to go out for the fresh air. Nothing would happen! She was sure they would enjoy some Chinese food too!

Hope was returning from her long walk. She got to the main entrance of the apartment building, scanning the periphery for any strange movements. Then she laughed at herself for being a believer! Nothing was going to happen!

She squeezed the bag with Chinese food against her body and fumbled for the key in her handbag. It was almost 9 PM now. The parking area was full with empty cars. Slot number 37 was occupied...meaning that Sam was home. He was not going to be happy with her...but...it was what it was!

A white ford car pulled up across from her and she made haste to fit the key into the lock. She eyed the vehicle suspiciously just to make sure that the boogeyman was not inside it. Nope! No boogeyman! She sighed and concentrated on turning the key. The lock snapped open. She cast a glance across her shoulder. The four men were on top of her!

A tall powerfully built one palmed her mouth and pushed something hard against her spine. She imagined what it was! "Don't scream," he instructed her softly. She nodded her agreement and he let go off her mouth. "OK, now you lead the way to your friends!"

Hope was so nervous that she dropped the bag of Chinese food. One of the men picked it up. "Hmm?" he mused, "How delicious!"

The diminutive fellow took time to swallow a few mouthfuls before carelessly tossing the bag away. Hope froze, as if she was too weak to climb up the stairs. The tall man jammed the nozzle against her spine, "Go on!"

She shivered at his cold voice. Hope put one foot forward and labored on upstairs. They had warned her! That made her heart ten times heavier!

Up in the apartment, Sam had just entered the room. Icilda and Kim were sitting in the couch. As he came in, Kim said, *"Hope's not here!"* in an urgent tone.

The big man scanned the room and shook his head in frustration. He got the message. He pulled his gun, spun around and stepped back to the corridor. There she was! Just seven meters away from him...frozen on her feet, and terrified as hell! A tall man was holding Hope from behind, with a gun pointed at her temple.

Sam locked eyes with him. He was confident. The cop would not risk taking a shot with the woman in harm's way. The thug grinned and then made a big mistake. He was turning the nozzle towards Sam!

He did not expect what happened next. Sam's gun rose in a flash, and he pumped off one in that same instant. The bullet shaved hopes hair and plugged a hole above the bridge of the thug's nose. The injured man's hand slackened on the handle of his weapon, which fell to the ground. He went down slowly, his body supported by her. She fought desperately to free herself.

Sam shouted to her, "Hope! Get down!" but Hope was not listening. He saw three other men coming up the stairs. The big cop expected that there would be more of them. What he did not expect was the degree of Hope's hysteria. She was not going to make it easy! Following instructions was never her forte. Instead of going down, she was running to him, right in the line of fire!

Sam charged forward, shouldering her aside that she crashed against the wall and fell flat on her stomach. Then, the first shot was fired! There was no way he could take evasive action. The big man felt the scorching bullet rip into his chest. He looked down in horror, before refocusing and firing back. A hole appeared between the shooter's eyes, and he was dead on his feet. Sam leaned against the wall and slid to the ground, while Hope charged into the room, screaming for Icilda.

The last two men edged forward cautiously. They had gotten rid of the only threat, but they were not going to take any chances.

Icilda's head popped around the door. The little woman made a quick scan. She saw Sam down in the corner, and the attackers heading for the room. Her eyes shot open with rage, and she hissed like a defiant cat, "Raas hole! You shot my man!"

The attackers grinned, thinking she was funny. Then the one in front saw the shiny gun rise to his face and froze in wide-eyed terror. She caught him off-guard! The only defense he could manage was to scream a useless warning. "Lookout!"

Icilda fired once, hitting him in the stomach. He dropped his gun inadvertently, trying to hold up and quench the fire in his intestine. She walked over to him nonchalantly, oblivious to the fact that his crony was firing at her! The injured man's eyes were begging her for life. Icilda pumped two bullets into his face. The other man aimed again in a hurry, fired two more shots and missed. Icilda focused on him, taking her time. Her fearless attitude freaked him out! "Crazy ass fucker!" he hollered unconvincingly. He fired three more times in rapid succession, but it seemed like luck and fearlessness worked together. She was concentrating hard on her aim. He was missing...she was not! The diminutive man turned tails and made a dash for the stairs, just as she squeezed the trigger. The bullet hit the metal bar, exactly where he was standing! He jumped, skipping imaginary missiles and emitting stressful whimpers. He glanced back with desperate hope that she would quit, but the stubborn little woman was taking it to him like a hungry predator!

"Icilda!" Kim called, running out to Sam. She was a nurse from the lab, and that was hope to Icilda. The little woman hollered over her shoulder at Kim, "Call 911!"

Icilda took off after the shooter, panting like a pit-bull. Kim's body raised goose pimples on hearing the desperate thirst for blood in her friend's voice.

She could not make him get away! Sam told her not to make them get away! The man scampered down the steps with the little woman in hot desperate pursuit. He pushed the gates ajar and stopped suddenly! Then he took to reverse! His gun came loose from his grip and he slowly levered it to the ground.

Alvin had his .44 barrel pointed between the nervous man's eyes. Icilda flew down the stairs in her nightdress, bare-feet, with a big chrome 9mm pistol in her hand...and it was red hot! Alvin's jaw dropped! The little villain was overly ashamed, especially after seeing Alvin's flabbergasted expression.

Then the tears started down Icilda's face. "Alvin! They shot Sam!"

"*What?*"

....

Halloway was at Marilyn's place when Alvin called him. "What's up pal?"

"We have a bad situation here!"

"How?"

"Faith Defender attack! Hope went out and attracted them! Sam got shot!"

"*Fuck*! How bad is it?"

Sam paused too long. Halloway was about to repeat the question when Alvin responded. "Dead Halloway...Sam's fuckin' dead!"

The big cop nodded understandingly. He was too dazed to realize that his colleague could not see him react to the information.

"Halloway?"

He heard Alvin talking to him, but it seemed like a dream. "Yuh...I hear you..." Halloway replied hoarsely. "How are the girls?"

"Safe for now...but you need to come over. Get Charley Boy! We need to put an end to this shit!"

"I know! I'm thinking about something Marilyn said...Maybe it's a good idea."

"What's that?"

Marilyn was coming from the shower when she heard her name. She walked over beside Halloway, curious.

"She said the people want them dead, so we need to make them dead!"

Alvin stopped to think. "That makes sense! Where did she get the idea?"

"I don't know!"

Suddenly Alvin exclaimed, "Halloway! I forgot to tell you! We got one of the assholes here waiting for you to come! This one is ours – not for the Boston Boys! I'm sure he will help us a lot!"

"Don't let anyone know we have him."

"Just get over here fast!"

Halloway hang up the phone. Marilyn was reading his face. "Bad news about the girls?"

"It's more like bad news about Sam. He's dead! A Faith Defenders attack. One of those girls went out of line and drew the heat!"

Her knees felt weak! Marilyn dropped down beside him on the couch. He was rising in a hurry, and she rose again to tail him. "So what are you going to do now?"

"We'll *'make them dead'*! That's what you suggested – Right?"

"Good!" she cheered, "Finally!"

Marilyn suggested it to Halloway days before. Now Sam had to go and die before he decided to do it. She knew how well it worked! That strategy was the Aduna Gyasi specialty that was keeping her alive.

Halloway arrived at Sam's place at 11:30 PM. They brought the surviving attacker to Kim's house and torched it while he watched, trembling.

He was uncooperative until Alvin realized the trick. The most intimidating thing to this fellow was the little woman with the big chrome 9. While the fire blazed, Alvin went to find Icilda by the car. She came out, gun in hand and took the culprit's phone from the cop. "He's all yours!"

Icilda stepped to the diminutive figure that was standing between Sam and Charley, watching the fire with utter disbelief. These were the most crazy, unorthodox cops he had ever seen! They would actually let this mad woman kill him!

"Here!" Icilda offered, stretching his own phone to him. He took the instrument from the woman, but was looking at her as if he was dumb. "Make the raas call boy!" she snapped with her blood-lust tone. He watched the cops, hoping they were not about to go too far with her. They did not care!

He took the phone and dialed out. "Speaker phone!" Icilda warned, cold as ever. They picked up immediately, "Hey pal!"

"Hey boss...It's me, Jo-Jo!"

"I know Jo-Jo! Just tell me what I want to hear!"

The man took a deep tense breath. Icilda raised her nozzle. He swallowed and said, "It's done sir! All three of them...dead..."

"How?"

"They were killed in a fire! There house was burnt down with them trapped inside!"

"Oh hell! That would have been painful for them! Did they scream a lot?"

He eyed Icilda nervously. "Y-Yes sir...A little...I mean, a lot!"

The man on the other end paused and then said, "Good! But how are we to prove it?"

"Well, the news will confirm it sir. The place is crawling with cops now...and they found three bodies, which they'll identify soon...by DNA..."

"Fine! They'll know within the next hour then! You earned your money...but you'll receive the transfer in two...after I've satisfied myself!"

The line went dead and Halloway grabbed the phone from the little man. The number he called did not exist anymore! He only had a single call! Alvin was taping the sound of the voice with an analyzer, which was one of Aduna Gyasi's creations. Within minutes, the analyzer would identify the voice and match it to that person from any phone call he had made before that one. No match! In fact, the analyzer could not identify it as a voice at all. They were dealing with a powerful person who had access to the cutting edge of technology. A deposit would be made to Jo-Jo's account but they knew that it would be another dead end. Twenty million dollars was a good start for the girls though.

The person who paid these Faith Defenders rogues also paid for a lifetime of secrecy. With so much money at stake, the zealots would never stop coming forward. Thankfully, Gyasi was dead. It could not last forever.

Two hours later, Marilyn Trooper covered the news about the tragedy for Boston Day. Jo-Jo got his transfer and used his phone to redirect it to Kim's account. Icilda called her friend to confirm. It was there!

"Well!" Charley boy announced cheerily, "What are we waiting for? We're done here!"

"Yeah!" Alvin agreed. He took Icilda by the elbow and walked her back to the car. Halloway waited until they were meters away. Then he pushed the little man forward. "Move!"

Jo-Jo headed forward with the big man behind him. The little man was looking over his shoulders spitefully until he finally said, "You're all dirty cops! I'm gonna fuck you up for stealing my money!"

Icilda slowed to look behind her and Alvin hustled forward. "Come on Icilda! We've got to go!"

She turned, pointed her nose forward and stepped up the pace. Then Halloway made his move. He grabbed the little man from behind and slit his throat with his big knife! There was no way he could go telling everyone about the living dead women.

31

Five weeks later, Doo-Bug was pouring her morning coffee, while she concentrated on watching television. "How are you planning to spend your day off my lady?"

"Good question Doo-Bug! After such a hectic period, I really don't know!"

"Did you see Angel last night?"

She sighed, "No Doo-Bug! The last time he came, he said…"

"It would be the last?"

Anne Eve frowned, "Yeah! How'd you know?"

"Anne Eve, you're a goddess. Fairy tales are made of this stuff. Science from our god-creator is magic to the sheep! I'm not just here to teach you; I'm here to see you perform marvelous works too! Angel is a mere program in you! You saw him when you opened it by your own doing! I knew who you were before you were born. My god made me to serve you! The prophecies to me are that I should observe the signs. When he makes the last visit, you've opened the final program, and fulfilled that prophecy!"

Anne seemed confused. "If Angel is my internal program, why is he real in other people's reality?"

"Because you're real, and you create the hologram. You are superior. Be aware of your own consciousness and see the god in you!"

"With all these incredible things I hear about me, I still feel 'common'!"

"We should go to Shrewsbury my lady. There's something you should know!"

"About what?"

"Let's go! You'll see!"

She decided to make the trip that day. The problem was to get Halloway off her! Even though she told him that what happened between them was a mistake, he kept hounding her! She felt like she would throw up! Marilyn turned off her phone and went!

She spent the most wonderful day with Dianne, who was excited to see her. Still, she was 'anticipating'. Doo-Bug said there was something for her to 'know'!

Evening came and no one told her anything. At that point, Dianne asked, "How are you coming on with the project?"

"Mother, is that a trick question? How can I do something that I don't know?"

"Are you sure you don't know? You are the project."

Anne stopped to think. She remembered that night with Angel and her last conversation with Doo-Bug. She got it! Was this the lesson? She would not tell her what she did with Angel though! "I am being what I am the way I was created to be."

Dianne nodded, "The project is safe in you!"

Apart from the joy of being with Dianne, she was disappointed! Doo-Bug raised her expectations before she came! When she was about to leave in the evening, the woman took her to her room upstairs. "There's someone who wants to see you!"

"Angel?" she asked, knowing how impossible it was. Then she remembered! Aduna told her that she would meet 'someone' when the time was right! Dianne pushed the door open. Aduna Gyasi sat at his desk working as usual! A robot that looked much different from Doo-Bug stood by his side. He turned his head and smiled all-knowingly. "Come on in my loveliness! Meet Melvin!"

Marilyn froze, not knowing whether to expose her anger or her joy first. Both needed ventilating! "Why?" she scolded.

Gyasi pointed to her stomach, which was not showing! "There is your answer! Nothing would go on if you did not go and do something!"

"How'd you know what I did?"

"I know before you were born!"

Anne changed the topic. "I saw you die, and it was *you*!"

"That thing was not me! Doo-Bug told you! You did not listen and you did not look! The consequence to you was pain! I planted just enough of my memories into it."

"I see," Anne croaked, disappointed. She wanted him to experience that special moment when she kissed him on the cheek! It never was!

"You must start seeing the simple things. Aduna Gyasi is never wrong!"

Now she realized! It was a part of her training! This was not the end of anything. It was the beginning. Aduna rose for her and she went to his embrace! "I'm so happy Father! I love you too much!"

"You would love your brother too if he were here?"

"I love Angel like I love me Father!"

"Oh…" the old wizard hesitated, "You learn something at a higher degree and then you reason it at a lower one. You cannot continue like this my dear. When you learn to listen, you will be fine!"

"I missed the point again, did I?"

"Do you not always?"

That night as she got home, she called Halloway, the man who did not kill her father but tried to do it anyway!

He picked up on the first ring, "Marilyn! I've been calling you!"

"I turned off my phone!"

"Special assignment?"

"No...but I had a lot to think about."

"Can I come over?"

"Please, let's not get into that again! Halloway…I'm pregnant!"

He paused and her heart started to race. Then he exploded on the other side, "What? That's great news to me Marilyn! Please have this baby!"

"I'm already having it! That's why I called you!"

"I'm one hundred percent in support!"

"You better be!"

He was screaming in ecstasy even as she hung up the phone. It disgusted her! Halloway was a dog that wanted to 'pass its place'!

....

Six weeks later…

After the destruction of Aduna Gyasi's project, the authorities declared that it was safe for women and children again. The Faith Defenders put out a press release, stating that they were satisfied that evil was defeated by the hands of faith. They warned that they would be ready to take up the challenge again, if necessary.

Tutu Babe was watching 'Good Morning with Marilyn Trooper'. Adele, Marilyn's co-host poked fun at her. "Let's see if our viewers can guess who is going to have a baby!"

It was easy! Marilyn was *showing*! Babe did all he could to destroy Gyasi's project, even to the death of the wizard. Aduna kept Marilyn Trooper at the East Lab for more than a year! He kidnapped her and there was no telling what he could have done to her. Besides, Marilyn Trooper lived like a nun! That was always reason for conversation. She was a good kid, but he would not take chances. He had to order the hit!

Tutu lifted the phone to call someone, who would want to make money on that. A thought struck him and he dialed the editor's number instead. Marilyn Trooper reminded him of Dianne!

"Hello?"

"Hey! Tony Patterson! My favorite friend!"

Tutu heard the uncertain pause at the end of the line and chuckled, enjoying Tony's fear. "What is it now...Sir?"

"Tony! I'm disappointed in you! You think I have to be asking for a favor to call you? I feel so let down!"

Tony did not answer him. He just sat there looking through his office window and waiting for the devil to state his cause. Finally. "Tony, your best girl is knocked up. Isn't it suspicious?"

"Why?" the journalist asked hoarsely. "It is a personal matter for her anyway. I don't let things like that bother me Mr. Babe."

"You got me wrong pal! I mean, considering that she was a captive in the lab for more than one year. Couldn't it be possible that Aduna Gyasi did one on her?"

"Sir, respectfully, I don't see the logic there. She got pregnant after escaping."

"How'd you know that? I mean, Marilyn Trooper never had male friends..."

"She does now!"

"Since when?"

"Detective Halloway! Call it gratitude or whatever you like. He's been balling her from the time he brought her back to civilization!"

"What a waste! And what is he saying about..."

"Her pregnancy? That's all he talks about! Guy's been keeping a party every week in celebration!"

"He has to! Got more than he deserved! I don't think he had a kid either!"

"No. I think they solved each other's problem."

"*Ewe*! I'll see you around Tony!"

Tutu hung up the phone and leaned back in his chair with his arms holding up his neck. He had done what he needed to do. The world was back in line. Everybody could have babies again, as long as he got his girls and fetuses! The Faith Defenders did their work. Now he was ready to take supremacy. Nothing or no one on earth would challenge the new and improved Iron Soldier. The government was set to go into his hands!

32

November 18, 2027. He was born! Marilyn named him Angel Ashanti Halloway. He was 'Angel', for he came as an agent of the most high, Aduna. There was going to be a great future war. At that time, the demigods would need soldiers. He would be the father of legions. Arpago came because of that war; and that was why she called him 'Ashanti'. Why was he *Halloway*? That was a technicality! She was father/mother to a Gyasi demigod! The undesirable human that offered the name was a mere utility!

She delivered the child before Halloway arrived at the hospital, a bouncing boy at over nine pounds. The doctor handed the newborn to her and indicated that they could bring the father in briefly. Halloway charged into the room, his hand trembling as he reached for the baby. "Wow! He's as firm as a five-year old!"

Marilyn watched him, thinking about the value he was adding to the project. She hated him but she was compelled to reward him with a warm smile. "I owe you one John!"

He went to his knees beaming with pride. Then he leaned over and kissed her cheek. She indulged him gracefully. The true meaning of her words was lost to him. Humans were dumb! Not even their detectives could figure! How long could she tolerate him around her and her child? She remembered her father's words, and now they held greater meaning than before! *"You are sufficient in yourself. Time will prove it! A man who touches the daughter of a god to profane her sanctuary must surely die! A woman who bears fruit for the son of a god shall be raised above the children of men! She will feed at the gods' table, for she was mother of a god, who suckled him!"*

....

Three months later.

They were raising the Iron Soldier and she went to
the press briefing. It was her first job since giving birth.
She stood in Tutu Babe's Great Conference Room to see
the wonder of the day. They said they fixed the glitches.
Babe assured the media that there was no chance of
a malfunction this time. To prove his confidence, he
would be present at the awakening. They made a new
one from scratch, claiming it was better than the old one.

Eric pushed forward with his camera trained on
Marilyn, hoping to reach the best vantage point. He
wanted to capitalize on an open opportunity to get closer
to the podium. She stopped, seemingly lost in thought.

After the Gyasi experience, she lost the ability to
relate to the vanities of Earth. She had nothing to
contribute. All she could do here was dishonor. The
media was a propaganda machine for people like Tutu
Babe and children of the Watchers. It betrayed the
people, but the fools believed in them. Believing was all
they learned to do in their cultures of bondage! They
depended not on their own understanding but took ready
done definitions and prescribed notions of norm from
shepherds and definers.

'The Day' expected Marilyn to ask hypocritical
questions too, more to endorse than to inform. She
was Gyasi's daughter! What was there to ask of these
common people? Another person now stood in the place
of Marilyn Trooper. Anne Eve Gyasi did not belong to
their paradigm!

"*Marilyn?*" Eric poked, waving his palm at her face
from behind the camera.

"Huh?"

"You OK?"

She cleared her throat to assure him, "Yes Eric. I'll be fine. I'm back on the planet!"

Eric chuckled and waved her forward.

These humans were playing with fire! She sensed fear inside the room, which made her uncomfortable. It was good to know that the great Aduna Gyasi had her back! Even in 2028, people still imagined a god up in the clouds! She knew that her mighty father, watched over her from his secret place, as the great god-creator he was.

At 10:45, Simon Muir pressed the button. The witnesses stiffened involuntarily, suppressing their common need to panic. A broad grin crossed Simon's face. He could not resist causing them the embarrassment! The little man beamed confidence, which helped the fearful ones to believe. Tutu Babe stood beside the Iron Soldier offering a reassuring smile to the audience. That helped too!

A smooth hum of robotic energy ignited in the conference area. The Legion's eyes opened. It ran through the color spectrum excluding red, until it settled down on blue. *Blue* was good! Marilyn breathed easily and smiled, musing over the unified sound of exhalation from the people.

The media regained their composure! The questions began to rain! Tutu Babe left a senior scientist named Rodman to answer them. The billionaire took off for his office, waving Simon Muir forward. Simon took up his cue behind the big man.

Inside the office, Tutu Babe made himself comfortable, hoisting his feet upon his desk and waiting for a timid looking Simon to sit down. The little man sat and watched the boss questioningly.

"Simon Muir, my man!"

Simon did not answer. Babe continued. "I won't beat around the bush! I know you came from the best school. That's why I'm happy to have you here!"

"Thanks sir! I was lucky to learn from the greatest!"

"I know about you. That 'luck' you talked about was no luck at all! I instructed James Carter to introduce you to Aduna Gyasi! Just as I did for the last ungrateful reject! Now it is time for you to repay the favor!"

Simon shifted uneasily, a confused disbelieving look on his face. "*What* Sir? I didn't know you…"

Tutu Babe ignored his response, "As you can see, the world is almost in my hands now! Before I launch the Iron Soldier, I need you to come forward with the finished device. We will not tolerate failure from our scientists this time; especially seeing that there is no Aduna Gyasi around to counteract us in this episode!"

"I shall do my best sir! I am getting close to calibrating the machine so we can equip our beast."

"I really hope so! Doctor Uche told me he had it right, but when we tried it out on the Ivory Coast, it didn't kill a single nigger down there! I could not help thinking that, being an African himself, he did not want to kill his own! You're Caucasian. I expect much better from you!"

Simon swallowed hard. "I know how to fix your machine. Aduna Gyasi taught me well! I also know that collateral damage is necessary in war and science!"

"I don't know what you know. All I know is that I need it done the way I want!"

"I understand sir. You will get what you want."

"I see! Uche told me the same thing too! Did you know he had an accident?"

Simon scratched his head nervously. "I heard he fell from the tenth floor."

"You heard right! Even before the accident, he tricked us again! That's why we're still here fucking around for answers!" Tutu agitated. He changed the subject. "The first time we activated the Iron Soldier, they thought it malfunctioned. Even Aduna Gyasi believed that crap! It seemed his predictions happened. My approach was to create chaos! Humanity having to fight a machine that went haywire was a good script. At the same time, I would take out their security forces and systems. I underestimated Aduna Gyasi's genius! If we knew what he knew, we'd kill him a long time ago. In the end, we had to kill him with the knowledge! No one on earth today is anything like he was. I'll not unleash the beast at this point. I'll wait for you to give me what I want. I have the switch, just like the last time! I'm the only one who can control and order the machine. Those who can't are my subjects! I'll have use for you then Simon, but it all depends on you!"

"I'll fix the device in five minutes Sir! And you're right. There's no human left on earth to challenge Tutu Babe Cooperation. I mean…show me the man!"

Meanwhile, at Marilyn's house…

The three-month old baby's eyes shot wide open. Doo-Bug stood in front of the child with an angry pit-bull baring its teeth at it. Little Angel watched innocently from twelve feet away. The robot unleashed the bloodthirsty animal and ordered, "Kill the little sucker! Make Tutu Babe happy!"

The animal dashed forward for the kill. In that instant, Angel opened his mouth to scream! Doo-Bug sensed a slight vibration going through its robotic frame, but it could hear no sound. In a split second, the dog fell dead and its body skidded all the way to rest against Angel's stroller.

"No fucking way!" Doo-Bug mused. Angel lifted a knuckle to his face and yawned. Doo-Bug countered, "I'm boring you? A demigod wants to challenge a robot! You're greater than Gyasi, not greater than Doo-Bug! Let's see you do the next assignment!"

The robot walked over and toe poked the animal. It was dead. Blood trickled from its ears and nostrils. "I'll give it to you! You messed him up real bad!"

Doo-Bug turned to check on the other two animals chained to the wall. They were watching undisturbed, as if oblivious to what had happened to their counterpart. "Great balls of Aduna! It was loud, but only the target one heard it! Robots are in trouble! Demigods are coming! I babysit an alien child that gives my robot ass the creeps!

33

On April 1, 2027, Halloway brought three DNA samples to Tutu Babe's mansion!

The super cop sat in the big brown leather couch waiting for the billionaire to come out. After fifteen minutes, a tall skinny Tutu Babe came into the living room with two beautiful women on either side. He waved them off and approached the cop, bearing a state of the art sample analyzer in his hand.

"Super cop John Halloway in person! It's a pleasure to finally meet you!"

John stood up and reached for the handshake. "The pleasure is all mine Mr. Babes! And considering the amount of money you're paying me for nothing, it gives me even more pleasure!"

Tutu Babe chuckled. "I understand! Nevertheless, don't think I'm stupid. I just love being certain – that's all!"

"I understand! Just like I love getting paid!"

"Right!"

Tutu Babe rested the device on the coffee table and took a seat across from the cop. He sparked up an electronic Cuba Cigar and drawled, "You don't smoke?"

"Naw! That was a long time ago! Coffee works fine!"

"How'd you like it?"

"Straight black! Coffee and water!"

Tutu Babe pressed the intercom on the table, "Hey there Albert! Tell Jane to bring some black coffee for my friend over here!" He turned his attention to Halloway and beamed, "Well let me see what you got! Either way, you'll take me out of my misery for good!"

"There's no either way Mr. Babe! You're paying me a million dollars to prove to you that my son is my son!" He handed over the bag that rested on his lap, "Here!"

"Got 'em all?"

"Sure! You think I'm stupid? I'm serious when it comes to making money! One for me, one for Marilyn and the other for Angel."

Tutu Babe smiled, "'Angel'? That's what you call your kid?"

"Sure! *Why*?"

"Nothing...I just thought it was a nice name." Babe put his hand in his robe pocket fumbling for something. He came up with a little bag and dropped it on the table beside the device. "Do you know what this is partner?"

"No. Enlighten me!"

The entrepreneur was about to speak but Jane came to serve the coffee for Halloway and he paused, watching her. She wore a mini skirt and when she bent over to pour the beverage, it was much revealing. "See that ass?" Babe questioned excitedly.

"*Huh*?"

"Do you see that ass?"

Jane turned around watching Halloway as if to test how bold he was. "Yes I see it!" the cop replied casually, and Jane surprised him with a warm smile.

"Go ahead!" Tutu Babe poked, "Take a slap!"

Jane finished pouring and was about to go. She paused. "*What?*" Halloway quizzed.

"Slap it! Have some fun! You work too doggone hard!"

Jane took a reverse to him. Halloway watched childishly. Finally, he said, "Alright! I think I just might!"

He slapped her and she walked of giggling. "See?" Babe asked like a schoolteacher, "That wasn't so hard – Was it? It's one of the things that money can buy!"

"I guess!"

"OK now! Open the samples and pour them all in!"

"You can't do that! How will…"

"Don't worry! Aduna Gyasi made this analyzer himself! You'll see!"

Each time they put in a sample the device asked for the name of the contributor. Halloway punched in the name for each sample. Babe put in his sample and typed in 'Aduna Gyasi'. Then he turned the device to scan and analyze. After three minutes, it printed a page: *Aduna Gyasi – Progenator*

Marilyn +Angel

John = No relations

John read the thing again in disbelief. Tutu Babe was not surprised. "We've just found the cyber baby Halloway! What're you thinking now!"

"I'm thinking that I'll kill it myself...and *her* too!"

"Well please do! The god being must die, and *she* must die. They're a threat to us humans. Make it happen and multiply your one million a hundred times!"

Halloway was overwhelmed with the pain of jealousy comingled with the rage of betrayal. From that moment on, he knew that it would not dissipate in a hundred years!

"My god! I never hated anyone so badly in my life!"

"Instant hate! Tell me about it!"

Tutu Babe watched him grimace in pain and said, "Don't take this information to the security forces, because your integrity will come into question! We will do it as quietly as possible. Give me their location and I'll send people over to 'fix' it."

"No! I'll kill that abominable kid myself!"

"And what about *her*? The price is fifty million! After what she put you through, I bet you want to make it a hundred...I mean, fifty for the bastard too!"

Halloway stood up and sat back down thinking. Finally, he said, "Mr. Babe, I was a fool with my affection! Money can by love when I want it! I will kill Marilyn Trooper and I will kill the demon child!"

....

Her father needed Doo-Bug that day, to copy some of its programs to Melvin. Otherwise, he would have to do the entire process from scratch.

"Father, I don't feel safe without Doo-Bug!"

"Don't worry! Your brother will be watching!"

"I told you Angel is no longer with me!"

"Let us not argue, for you never listen!"

"I never listen because I'm Aduna Gyasi's daughter!"

"What about the fact that you do not try to be rational?"

"That part is because you dumped me for twenty-eight years!"

"Nothing ever happens Anne! Why would it be different today?"

Doo-Bug left and she went upstairs for Angel. At least she would spend time with him alone. "Today," she told him affectionately, "we hang out on the couch together in my living room!"

The day went quickly, as she was caught up in the joys of Angel. It was almost ten at night! Doo-Bug had not returned!

Immediately, there was a knock on her door. "Think of the devil!" she quipped, rising in a hurry to open it. Halloway stood in front of her! Usually he never came over without telling her, and she felt funny about this visit. Her heart warned her to close the door, but she pushed it wide open and stepped aside. "What is it?" she asked, fighting not to sound agitated.

"I came to see you and the baby! Is something wrong with that?"

She thought about it. He was never so brazen before. Under the circumstances, she gave him a right to plead, not to demand! "I didn't say that!"

"You didn't say that!" He reiterated sarcastically, "What about sex?"

"*What?*"

"What about when I want to…"

"*Why*? It doesn't happen Halloway! It doesn't happen at all!"

"You were the one who came on to me!" he hissed demoniacally.

His last words caused fear in her. Her temples throbbed with each beat of her heart! "What're you here for John?" she asked uneasily.

"This!" he snarled, grabbing her wrist and twisting it behind her back. Then he forced her into the kitchen. Marilyn did all she could to resist, but he was strong as an ox! He pushed her up against the kitchen counter, applying painful pressure against her arm. She felt his repulsive touch as he fumbled to remove her underwear! She had to do something! He had the physical ascendancy. All she had was her mind! Her left palm was resting on the marble counter and she concentrated on the energy it contained. It charged her body and she pushed hard against him, throwing him off her.

Halloway fell on his back, looking awestruck. "You're a strong bitch aren't you?"

She glanced toward the living room wondering if she could get past him to grab her baby and run for safety. Her action telegraphed her thoughts and he smiled demoniacally. "I'm calling the police!" she threatened.

"Go ahead!" he chided, "They'll protect Aduna Gyasi's daughter and grandchild!"

"What're you saying?" Marilyn asked, shocked.

"I know the child is not mine, and I know the state wanted it dead! They would sentence you to execution if they found out Marilyn!"

She put out her hands, knowing it would not make sense to deny it. "We're innocent Halloway!" she pleaded. "Can't you see we're who we are?"

"You lied to me – Whore!"

"I lied to survive! Don't you see?"

"I came to kill you!" he informed chirpily, rising to his feet with his pistol in hand. The child sat up like a one year old. Its eyes turned to red. But, the lock turned on the door and it went back to sleep.

A white bearded fellow in white walked in. It shocked the big cop, especially as he did lock the door. "Who the fuck is you Alibaba?"

"*Alibaba*? If you think you know who I am, why ask me? I'm a child of gods! Who are you? Barbaric man?"

In a sudden flash, Halloway fired, but something strange happened. The little man with the girlish voice had disappeared! The big cop spun around confused. His eyes caught Marilyn's, questioning her. She shook her head to say that she did not know him. Then Anne saw the stranger standing behind the big cop!

Halloway spun around again and fired. Marilyn charged from the kitchen in fear. She grabbed Angel and went through the door. Someone else was out there! She stepped back inside, closed it and headed for her bedroom. She locked herself in and sat on the side of her bed, holding onto Angel and trying to think. From down stairs, she heard a few more shots followed by silence. Then she heard yet another shot.

Halloway was turning on his heals looking for his target again. "What the fuck kind of a creature are you though? Are you one of those cyber things?"

"I am a god to you," the sweet voice whispered from behind him. Halloway felt a touch on his shoulder and pain shot through his body. "Aaagh!" the super cop screamed.

Anne heard the scream, but was not sure whom it came from. The stairs creaked. Someone was coming up to her!

"Great balls of Aduna!" she muttered, "I should've called our father!"

Angel touched her. She watched him quizzically. He pointed to the door! Anne frowned. She had no idea of her own, so she went to open it. The stranger stood on the steps in front of her. "Hello little sister! Meet my friends Alexia and her baby Marie!"

Marilyn frowned. Angel pulled eagerly towards the stranger and she realized the connection. "Is this my nephew that the greats prophesied about?"

"Is that your daughter?" Marilyn asked, pointing to the child in Alexia's hand.

"No," the little man informed her. "She's a daughter of men, but she will learn like the gods do! And this, her mother, is now as a sister to you!"

"You're married?"

"What is marriage? I have chosen her!"

"I see…"

"So, are you going to pack and come with me now?"

"Did you kill him my brother?"

"Who – that big man who wanted to kill you? No, I didn't think he was worth it!"

Marilyn stomped down the stairs. "He has to die!"

She handed Angel to Ray and headed for the kitchen. When she returned with the big knife, she saw that the detective was gone! Ray watched her, amused.

"My brother…whatever your name is…"

"Ray. Don't you know about your family?"

"Where's he?"

"He scampered off when he heard knives rustling! Don't be a scaredy cat though!"

"He'll come back…to kill us!"

"Tonight he won't. He will by tomorrow though! Let's have you pack before then! He left his gun on the floor! It's for you!"

Marilyn walked back past him, eying Alexia as if seeing her for the first time. The ginger head woman reached timidly for a handshake. Marilyn refused, opting instead to take the woman and her child and pull them to her bosom.

The End